AS / A2

OCR
Revise
Human Biology

...tter support for you

Barbara Geatrell, Sue Hocking
and Jennifer Wakefield-Warren
Series editor: Fran Fuller

w.heinemann.co.uk

✓ Free online support
✓ Useful weblinks
✓ 24 hour online ordering

0845 630 22 22

D0258830

OCR
RECOGNISING ACHIEVEMENT   Heinemann

Official Publisher Partnership

Heinemann is an imprint of Pearson Education Limited, a company incorporated in England and Wales, having its registered office at Edinburgh Gate, Harlow, Essex, CM20 2JE. Registered company number: 872828

www.heinemann.co.uk

Heinemann is a registered trademark of Pearson Education Limited

Text © Pearson Education Limited 2009

First published 2009

12 11 10 09
10 9 8 7 6 5 4 3 2 1

British Library Cataloguing in Publication Data is available from the British Library on request.

ISBN 978 0 435692 42 1

Edited by Priscilla Goldby
Designed, produced, illustrated and typeset by 𝒯 Tek-Art, Crawley Down, West Sussex
Original illustrations © Pearson Education Limited 2009
Cover photo/illustration © Science Photo Library/Neil Borden
Printed in China (CTPS/01)

Every effort has been made to contact copyright holders of material reproduced in this book. Any omissions will be rectified in subsequent printings if notice is given to the publishers.

Contents

Contents

Introduction

How to use this Human Biology revision guide

This revision guide is for OCR GCE Human Biology course, covering material from the taught units of both the AS and the A2 courses. Here is a plan of the units and modules you will study.

Unit	Module	Length of exam	Percentage of total GCE mark
F221: Molecules, blood and gas exchange	Module 1: Molecules and blood	1 hr written exam	15%
	Module 2: Circulatory and gas exchange systems		
F222: Growth, development and disease	Module 1: The developing cell	1 hr 45 mins written exam This exam has pre-release material. The first two questions are based on this.	25%
	Module 2: The developing individual		
	Module 3: Infectious disease		
	Module 4: Non-infectious disease		
F223: Practical skills in human biology	Coursework tasks that will be set by OCR and marked by your teachers		10%
F224: Energy, reproduction and populations	Module 1: Energy and respiration	1 hr written exam	15%
	Module 2: Human reproduction and populations		
F225: Genetics, control and ageing	Module 1: Genetics in the 21st century	1 hr 45 mins written exam	25%
	Module 2: The nervous system		
	Module 3: Homeostasis		
	Module 4: The third age		
F226: Extended investigation in human biology	A single investigation that involves planning, collecting and presenting data, and analysing and evaluating your results		10%

When you are revising, have a copy of the specification to hand. It is written as learning outcomes that tell you exactly what you should learn. As you revise, tick-off each learning outcome when you understand what it means, and can write about and explain it.

This book follows the sequence of learning outcomes in the specification. Each module is divided into a number of spreads, which end with some **quick check questions** to test your recall and understanding.

QUICK CHECK QUESTIONS

It is very important that you know the meanings of all the terms in the learning outcomes. These terms, and some others, are highlighted in bold in the text

Throughout the book you will also see boxes containing the **Key words** you need to know for your exam and helpful **Examiner tips** from experienced human biology examiners

Key words   Examiner tip

At the end of each unit there are **End-of-unit questions**, which resemble the types of question you will find in the examination papers. **Answers** to all of the questions are on pages 180 to 199.

It is most important that you know the meanings of all the terms in the learning outcomes. These terms, and some others, are highlighted in bold in the text.

As part of your revision, you may wish to follow up some of the ideas in the course, especially in Unit 2 (F222 which uses case studies. One good source of information – among many books and websites available – is www.askabiologist.org.uk.

Topic		Pages in this book	Learning outcomes from the specification	Useful GCSE/AS revision
Unit 1 (F221): Molecules, Blood and Gas Exchange				
Module 1: Molecules and Blood	The Blood	2–7	1.1.1(a–j)	Calculating sizes from diagrams, prokaryotic cells (2.3.1c), Specialised cells (1.2.3b,c).
	Molecules	6–15	1.1.2(a–w)	Chemical symbols from GCSE diet (2.2.1e, 2.2.2b, 2.4.1h); 2.4.3d – testing for blood glucose; Practical work on osmosis in erythrocytes and plant cells.
	Preventing Blood Loss	16–17	1.1.3(a–e)	The pH scale from GCSE, protein structure (1.1.2c). Practical work on the effect of enzyme and substrate concentrations on reaction rates. 2.4.1f – how aspirin works.
	Blood for Medical Use	16–19	1.1.4(a–f)	Practical work on the effect of temperature and pH on enzyme activity. HIV transmission (2.3.1e).
Module 2: Circulatory and Gas Exchange Systems	The Heart and Monitoring Heart Function	20–21	1.2.1(a–f)	Calculations of heart rate from graphs. Experiments on the effect of exercise on heart rate. CHD (2.4.1b,f).
	The Circulatory System	22–23	1.2.2(a–h)	Using a sphygmomanometer; Using a microscope to examine arteries and veins; Calculations of sizes from photomicrographs and diagrams.
	The Lungs and Investigating Lung Function	24–25	1.2.3(a–k)	Review cell structure (1.1.1f); Using a microscope to examine tissues; Practical work using the spirometer and/or peak flow meter; Link back to 1.1.2d and the role of oxygen; Calculations of surface area to volume ratios.
End-of-unit questions		26–29		
Unit 2 (F222): Growth, Development and Disease				
Module 1: The Developing Cell	Mitosis as Part of the Cell Cycle	30–37	2.1.1(a–l)	Link to cell structure (1.1.1f) and cell specialisation (1.2.3c) as examples of differentiation. Practical work on staining dividing cells.
	Detecting and Treating Cancer	38–41	2.1.2(a–g)	From GCSE, types of radiation; Analysis of data from tables and graphs describe trend and patterns or to show correlations; 2.1.2d – link prevalence to 2.3.1b (epidemic and endemic) 2.3.3a (morbidity and mortality).
Module 2: The Developing Individual	The Biological Basis of Individuality and the Monitoring of Fetal Development	42–49	2.2.1(a–j)	Review mitosis (2.1.1a). Review haemoglobin (1.1.2c – iron in prosthetic group) and blood clotting (1.1.3b – the role of calcium). Review ultrasound (2.1.2c). Link 2.2.1g to oxygen transport (1.1.2d). Link antenatal care with blood pressure (1.2.2f,g).
	The Developing Infant	50–53	2.2.2(a–g)	Use growth charts to describe patterns of growth. Practise calculations of absolute and relative growth rates. Link to back to organ systems (1.2.3b), and 1.2.1 and 1.2.2 (the circulatory system).
Module 3: Infectious Disease	Controlling the Spread of Infectious Disease	54–59	2.3.1(a–k)	Compare prokaryotic and eukaryotic cells (1.1.1f) – remember to include plant cells (link to 2.3.1i); Link 1.1.4f – the screening of blood for HIV; Link TB transmission to 1.2.3b; 2.3.3a – epidemiological data link also here.
	Acquiring Immunity	60–67	2.3.2(a–m)	Link to 1.1.1j – organelles involved in secretion of antibodies. Link antibody structure to 1.1.2c (protein structure). Link ABO and Rhesus blood groups to 2.2.1d.
	The Future of Infectious Disease Control	68–71	2.3.3(a–e)	Link 2.3.1g – new strains. Also link to viral structure 2.3.1d for position of antigens.
Module 4: Non-infectious Disease	Coronary Heart Disease (CHD)	72–75	2.4.1(a–l)	Link to 1.1.2u for fat structure and 2.2.1e – for DRV values. Link to 1.2.1 for heart structure and ECG. Link aspirin treatment with enzymes and blood clotting (1.1.3d), and 1.1.4c
	Lung Disease	76–77	2.4.2(a–c)	Link smoking to 2.1.2b,c. Review the structure of the respiratory system (1.2.3).
	Diabetes	78–81	2.4.2(a–e)	Review testing for glucose (reducing sugar) using Benedict's reagent from GCSE. Link insulin to protein structure, and to glucose and glycogen (1.1.2q,r,s).
End-of-unit questions		82–87		

Unit 4 (F224): Energy, Reproduction and Populations				
Module 1: Energy and Respiration	Respiration	88–91	4.1.1(a–r)	Review oxidation and reduction from GCSE as the loss or gain of electrons (or hydrogen). Link to active transport (1.1.2j) and mitochondria structure (1.1.1f). Review structure of fats and proteins (1.1.2u and 1.1.2b). Practical work on respiration in yeast. Review factors affecting enzyme activity (1.1.3e and 1.1.4b).
	Athletic Performance	92–99	4.1.2(a–o)	Review 1.2.1e,f. Review DNA structure (2.1.1b) and haemoglobin structure (1.1.2c). Review erythrocyte structure (1.1.1d) and link to sickle cell. Review cell counting (1.1.1c) and link to EPO. Link DNA repair to 2.1.1b, c and d, and to 2.1.2a.
Module 2: Human Reproduction and Populations	Fertility and Contraception	100–105	4.2.1(a–k)	Calculations of size or magnification from diagrams or photomicrographs. Review meiosis (2.2.1a,b); Review antenatal care (2.2.1d,e,f); Review differentiation (2.1.1l) and link to sperm and oocyte.
	Assisted Reproduction	106–107	4.2.2(a–i)	Link to 2.3.2i (antibodies) for monoclonal antibodies and pregnancy testing.
	Food, Farming and Populations – Producing Food	108–113	4.2.3(a–l)	From GCSE review food chains and ecological pyramids. Review plant cell structure (1.1.1f) and respiration to explain energy losses (4.2.3h); Interpret data from energy flow diagrams; Calculations of increases in yield.
	Food, Farming and Populations – Human Impact on the Environment	110–113	4.2.4(a–g)	Review 2.3.3a. Review 2.3.1i,k (plants a source of medicines) . Calculations of percentage increases in populations.
End-of-unit questions		114–119		
Unit 5 (F225): Genetics, Control and Ageing				
Module 1: Genetics in the Twenty–First Century	Inheritance of Human Genetic Disease	120–127	5.1.1(a–l)	Review meiosis (2.2.1b,c). Review genetic code and protein synthesis (4.1.2e,f). Review karyotyping (2.2.1i) and blood smears (1.1.1a,b).
	Genetic Techniques	128–133	5.1.2(a–h)	Review DNA structure and replication (2.1.1a,b). Review bacterial cell structure (plasmids) 2.3.1c. Review protein synthesis (4.1.2e,f).
	Counselling Individuals on Genetic Issues	132–133	5.1.3(a–c)	Review inheritance patterns (5.1.1). Review contraception (4.2.1j,k) and assisted fertility (4.2.2a,b).
	Transplant Surgery and Cloning	134–135	5.1.4(a–e)	Review immunity (2.3.2), stem cells (2.1.1k,l), embryo storage (4.2.2d) and genetic techniques 5.1.2e).
Module 2: The Nervous System	Monitoring Visual Function	136–139	5.2.1(a–f)	Review eye structure from GCSE. Review Vitamin A (2.2.2b). Link to 5.2.2e,g.
	Treating Central Nervous System injuries	140–149	5.2.2(a–n)	Review 2.1.2c – detection of cancers. Review active transport , exocytosis and membrane structure (1.1.2j,k and 1.1.1h); Review stem cells (2.1.1k,l).
	Modifying Brain Function	150–151	5.2.3(a–e)	Review synapses (5 2.2g,h); Review ideas of specificity from 1.1.3d and 2.3.2i; Review 2.2.1g (alcohol).
Module 3: Homeostasis	The Importance of Homeostasis	152–155	5.3.1(a–h)	Review temperature and pH effects on enzymes (1.1.4b)
	Managing Type 1 and Type 2 Diabetes	156–157	5.3.2(a–h)	Review 2.4.3 and 1.1.2q,r,s. Linked to 5.2.1.
	Urine Production	158–163	5.3.3(a–i)	Review amino acid structure (1.1.2b) and 4.1.2h (Bohr shift). Review capillary structure (1.2.2a), active transport (1.1.2j) and osmosis (1.1.2i) Link to 5.3.1b.
	Treating Kidney Disease	162–165	5.3.4(a–i)	Linked to immune response (2.3.2d,j). Review EPO (4.1.2d) and 5.1.2d.
Module 4: The Third Age	The Effects of Ageing on the Reproductive System	166–167	5.4.1(a–f)	Review 4.2.1e and 5.3.1b.
	The Effects of Ageing on Other Body Systems	168–171	5.4.2(a–j)	Link to 5.2.2a,b, 5.2.1. See also 2.2.2b for vitamin D and calcium. Link to 2.1.2c (X-rays) and to 1.2.1, 1.2.2 and 1.2.3.
End-of-unit questions		172–175		

Blood

Key words

- leucocytes
- erythrocytes
- platelets
- neutrophils
- lymphocytes
- monocytes
- macrophages
- blood smear
- differential stain
- haemocytometer

✓ *Quick check 1*

Blood structure

Blood consists of plasma, containing dissolved solutes, and a number of different types of cell, each with a specific function. Blood cells are made in the bone marrow.

Leucocytes are white blood cells.

Key definition

White blood cells or leucocytes are involved in body defence. There are 7000 to 13 000 per mm^3 of blood, far fewer than red blood cells. There are a number of different types, each with a specific function; 72% are neutrophils (phagocytes), which engulf foreign proteins and pathogens; 24% are lymphocytes which produce antibodies.

Studies of these cells and their numbers may be used in medical diagnosis.

Structure and function

Blood cells	Size	Structure	Function
Erythrocytes (red blood cells)	7 µm	A biconcave disc Small and flexible shape No nucleus but packed with red pigment haemoglobin	Transport oxygen and carbon dioxide in reversible reactions Large surface area over which gases (O_2 and CO_2) diffuse quickly Easily change shape and squeeze through capillaries
Platelets		Just small amounts of cytoplasm and membrane with no nucleus	Release blood clotting factors on damage
Leucocytes:			
Neutrophils	9 µm	Have small granules in cytoplasm and lobed nucleus	Engulf microbes by phagocytosis
Lymphocytes	5.5 µm	Very large nucleus and little cytoplasm. Two types: B lymphocytes and T lymphocytes	B lymphocytes produce antibodies T lymphocytes destroy infected cells and have other functions
Monocytes	13–20 µm	Large cells with bean-shaped nucleus	Become macrophages after 3 days in blood system
Macrophages	15–20 µm	Large phagocytic cells in the blood, but also fixed within specific tissues, e.g. liver or lung	Engulf microbes and foreign cells in tissues

✓ *Quick check 2 and 3*

Hint

Remember platelets are not really cells but are **fragments** of very large cells called megakaryocytes.

Blood tests – taking a sample

Blood samples are taken from a vein in the arm or hand. The vein is made to stand out by applying a tourniquet above it. This makes it easier to insert the needle. An alcohol swab cleans the area and a sterile needle is used. Blood is sucked into the syringe and then cotton wool is applied as the needle is removed. Gentle pressure on this stops the bleeding and a plaster is applied.

The sample can be used for a variety of tests.

✓ *Quick check 4*

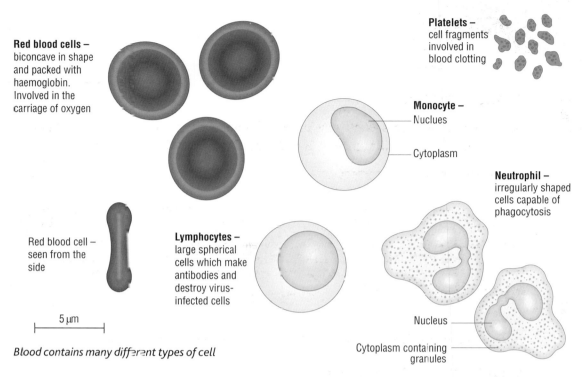

Red blood cells – biconcave in shape and packed with haemoglobin. Involved in the carriage of oxygen

Platelets – cell fragments involved in blood clotting

Monocyte –
Nuclues
Cytoplasm

Neutrophil – irregularly shaped cells capable of phagocytosis

Red blood cell – seen from the side

Lymphocytes – large spherical cells which make antibodies and destroy virus-infected cells

5 μm

Nucleus
Cytoplasm containing granules

Blood contains many different types of cell

Staining a blood smear

Part of the sample may be used to make slides for microscope observation.

- First a **blood smear** or thin film is made by spreading the blood across the slide using a second slide.
- The smear is labelled and allowed to dry.
- The cells are fixed using alcohol.
- A **differential stain** is added and left for 2 minutes (heat may be needed).
- The stain is washed off using water.

✔ *Quick check 5*

The number and appearance of the blood cells may help to identify certain conditions such as megaloblastic anaemia and sickle cell anaemia.

Using a haemocytometer

Blood cells may be counted by a laboratory machine or by using a special slide called a **haemocytometer**.

The slide is etched with a grid of known dimensions giving a total volume of 0.1 mm³. The blood is diluted accurately to a 1 in 200 dilution with a blood pipette. Cells are then counted in five of the 0.004 mm³ etched squares.

For white cells the dilution factor is 1 in 20. The diluting fluid causes the red cells to burst and so only the white cells remain to be counted. The larger squares on the corners of the grid are used, each with a volume of 0.1 mm³.

QUICK CHECK QUESTIONS

1 What might the number of white blood cells present in the blood tell you about the health of a person?

2 What are the important functions lymphocytes perform?

3 Why are red cells not found in tissue fluid but white cells are?

4 During blood sampling a number of sterile techniques are used. Why is this necessary?

5 Explain why a blood sample is smeared before it is stained.

Eukaryotic cells

Eukaryotic cells contain a nucleus. Simpler cells such as bacteria have no nucleus. All eukaryotic cells have specialised features in order to carry out specific functions.

You can see the **ultrastructure** or fine detail of the cell organelles under an electron microscope.

Key words

- eukaryotic
- ultrastructure
- organelle
- micrometre

✓ *Quick check 1*

Key definition

Eukaryotic cells contain a nucleus surrounded by a nuclear membrane and a number of cell organelles also surrounded by membranes. They contrast with prokaryotic cells which are more primitive and have no membrane-bound organelles or a true nucleus. All animal, plant, fungal and protoctistan cells have a true nucleus.

Leucocyte

There are several types of leucocyte (a specialised human cell) in the blood. Lymphocytes are one type that make antibodies.

Examiner tip

Learn the structure *and* function of all organelles in the diagram of the leucocyte. You should be able to identify these from any diagram or electron micrograph. Look at as many different diagrams as possible to help you.

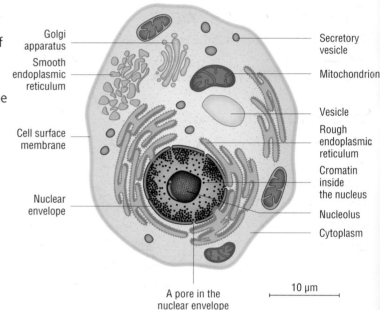

Ultrastructure of a lymphocyte

Organelles

The table gives the structure and function of the main **organelles** in eukaryotic cells.

Organelle	Key Structure	Function
Nucleus	Contains DNA packed as chromosomes Surrounded by nuclear envelope (a double membrane) containing pores	Carries genetic information Genes on the DNA code for proteins Controls the activities of the cell
Nucleolus	Dark stained area in nucleus	Production of RNA and ribosomes
Ribosome	Tiny organelle attached to endoplasmic reticulum or free in cytoplasm	Site of protein synthesis Assembles amino acids into proteins
Rough endoplasmic reticulum (RER)	Flattened membrane system within cytoplasm Has ribosomes attached to surface	Site of attachment for ribosomes Transports proteins to Golgi apparatus, by forming vesicles
Smooth ER	Membrane system without ribosomes attached	Produces and transports triglycerides, phospholipids and cholesterol
Golgi apparatus	A **stack** of flat membrane-bound sacs	Processes proteins by modifying them and packaging them into vesicles
Vesicles	Membrane-bound sacs budded off from RER and Golgi apparatus	Contain proteins made in the cell; transport them around the cell or to the cell surface membrane
Lysosome	Membrane sac containing digestive enzymes	Enzymes destroy old or damaged organelles, cells and engulfed bacteria
Mitochondria	Double membrane-bound organelles; much folded internal membrane with enzymes attached	Site of aerobic respiration and ATP production
Cell surface (plasma) membrane	Phospholipid bilayer around outside of cell	Controls substances entering and leaving the cell; partially permeable
Cytoskeleton	Protein microfilaments and microtubules in cytoplasm	Supports the cell and allows cell movement and movement of organelles within cells

Only plant cells:		
Cell wall	Outside the cell membrane, made of cellulose	Freely permeable Provides support to cell
Chloroplasts	Membrane-bound organelles Internal membranes form grana stacks	Site of photosynthesis Only present in **green** plant cells
Permanent vacuole	Surrounded by a tonoplast (the membrane around the vacuole)	Stores water and solutes to keep cell stable and turgid

A palisade mesophyll cell

Palisade mesophyll cells are an example of a plant cell specialised for photosynthesis and found in plant leaves. They are adapted for photosynthesis.

Interrelationship between organelles

In order to make complex proteins, cell organelles must work together. The diagram below shows how several organelles work together in producing and secreting proteins such as antibodies.

mRNA copies the DNA code for protein. Ribosomes use this to assemble proteins. A vesicle packages and transfers the protein to the Golgi for modification. Afterwards, they are repackaged in vesicles for transport to the plasma membrane. Mitochondria provide ATP.

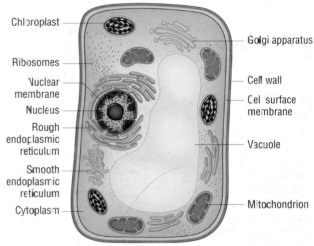

Ultrastructure of a palisade mesophyll cell

✔ *Quick check 2*

✔ *Quick check 3*

✔ *Quick check 4*

Examiner tip

Small structures under the microscope are measured in micrometres (one thousandth of a millimetre). Always measure the image (photomicrograph) in millimetres: just multiply by 1000 to convert to micrometres.

$$\text{mag.} = \frac{\text{image size}}{\text{actual size}}$$

Be prepared to calculate either the actual size or the magnification.

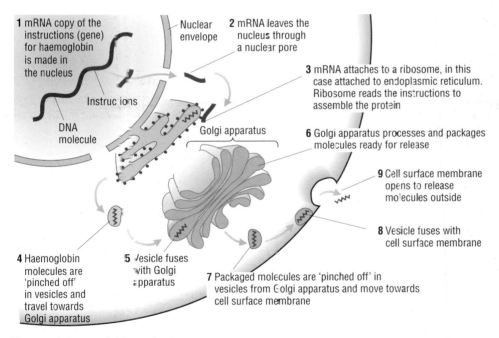

1 mRNA copy of the instructions (gene) for haemoglobin is made in the nucleus

Nuclear envelope

2 mRNA leaves the nucleus through a nuclear pore

Instructions

DNA molecule

Golgi apparatus

3 mRNA attaches to a ribosome, in this case attached to endoplasmic reticulum. Ribosome reads the instructions to assemble the protein

6 Golgi apparatus processes and packages molecules ready for release

9 Cell surface membrane opens to release molecules outside

8 Vesicle fuses with cell surface membrane

4 Haemoglobin molecules are 'pinched off' in vesicles and travel towards Golgi apparatus

5 Vesicle fuses with Golgi apparatus

7 Packaged molecules are 'pinched off' in vesicles from Golgi apparatus and move towards cell surface membrane

The protein haemoglobin production sequence

✔ *Quick check 5*

QUICK CHECK QUESTIONS

1 Describe the features of eukaryotic cells.
2 State the main differences between an animal and a plant cell.
3 Describe how organelles work together in an animal cell to produce a protein.

4 What is ATP used for?
5 What is a micrometre?

Membranes

All living cells are surrounded by a cell surface (plasma) membrane. Within the cell, membranes enclose or make up cell organelles separating the cell into compartments.

A fluid mosaic model

A membrane is:

- **fluid** – made up of phospholipids which are liquid and move easily within each layer
- **mosaic** – from the surface the membrane looks like a mosaic picture made of tiles. The proteins and phospholipids form the pattern.

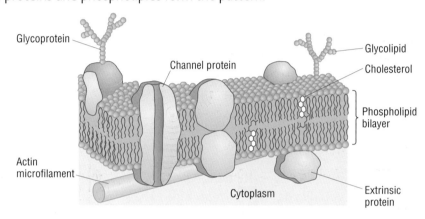

The fluid mosaic model

The membrane is a thin layer, about 7–8 nm wide, and made up of several types of molecules.

- **Phospholipids** are specialised lipids. They form a double layer (**bilayer**). They act as a barrier for water-soluble molecules between the cytoplasm and the outside. Their hydrocarbon tails are **hydrophobic** and point inwards. They cannot be in contact with water-based media like cytoplasm. They are held together by weak bonds. The phospholipid heads are **hydrophilic** and face outwards towards both the cytoplasm *and* the outside of the cell, both of which are water based.

Key definition

A bilayer is a double layer of phospholipids forming the basis of the fluid mosaic membrane. The phospholipids line up with the hydrophobic tails facing each other in the centre of the layer like the filling of a sandwich. The hydrophilic heads point outwards away from the centre. On one side they face the cytoplasm and on the other side they face the outside of the cell.

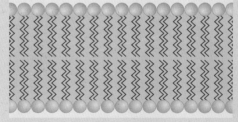

A phospholipid bilayer

- Proteins are part of the membrane. Large proteins extend across the membrane, smaller ones lie on the surface of the membrane.
- Carbohydrate chains are attached to some proteins (forming **glycoproteins**) or to some phospholipids (forming **glycolipids**).
- **Cholesterol** molecules are only found in eukaryotic cells.

Key words

- fluid mosaic
- phospholipid bilayer
- hydrophobic
- hydrophilic
- intrinsic proteins
- extrinsic proteins
- glycoproteins
- glycolipids
- cholesterol

✓*Quick check 1*

Hint

The cell surface membrane controls everything entering or leaving the cell.

Examiner tip

Learn the cell organelles covered in your course together with their functions.

Hint

Proteins extending across the membrane are fixed: **intrinsic** proteins. Others may float on the surface: these are **extrinsic** proteins.

Examiner tip

Glycoproteins and glycolipids are only found on the *outside* of the cell membrane.

Functions of the molecules in the cell surface membrane

Phospholipids:

- form a barrier for water-soluble substances between cytoplasm and extracellular fluid
- are permeable to non-polar molecules like fatty acids and oxygen, and to small polar molecules like water and carbon dioxide, and to alcohol
- are impermeable to ions and large polar molecules
- allow movement so vesicles can form or rejoin.

✔ *Quick check 2*

Proteins:

- act as channels or pores and as carrier molecules
- act as receptors for chemicals such as hormones
- act as enzymes in some membranes within cells.

✔ *Quick check 3*

Cholesterol:

- helps stabilise the membrane by binding the phospholipid polar heads and non-polar tails
- reduces permeability to water and non-polar molecules
- prevents phospholipids from solidifying at low temperatures or becoming too fluid at high temperatures.

Glycoproteins and glycolipids:

- act as receptors for cell signalling molecules such as hormones
- act as a recognition site (antigens) for cells such as leucocytes
- help to stick the cells together in a tissue (cell adhesion).

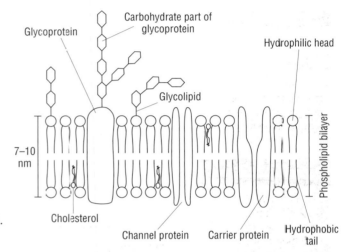

A cross-section through a cell surface membrane

Role of cell surface membranes

They:

- determine what is allowed through
- act as a barrier to many water-soluble molecules
- prevent large protein molecules, such as cytoplasm and enzymes, leaving cells
- are permeable to selected molecules
- allow larger molecules to pass through pores or via carriers, e.g. glucose
- or to enter or leave by endocytosis and exocytosis
- act as receptors or as recognition sites
- increase the surface area by folding into microvilli.

Examiner tip

You will need to draw *and* label the membrane structure.

✔ *Quick check 4*

Role of membranes *within* the cell

They:

- subdivide the cytoplasm allowing different cell functions to occur more efficiently
- provide a large surface area for enzyme or pigment attachment, such as mitochondria, chloroplasts or endoplasmic reticulum
- surround lysosomes which isolate harmful digestive enzymes
- surround vesicles which transport molecules between cell organelles or to the cell surface membrane.

✔ *Quick check 5*

QUICK CHECK QUESTIONS

1 Explain why the model for membranes is called fluid mosaic.
2 Describe the roles of proteins in the membrane.
3 Why is cholesterol important in membranes?
4 Explain the difference between microvilli and cilia.
5 Describe the importance of membranes within cells.

Proteins and haemoglobin

Key words

- a mino acids
- condensation
- peptide
- hydrolysis
- primary structure
- secondary structure
- tertiary structure
- quaternary structure
- denatured
- prosthetic group

Examiner tip

You need to know this basic structure of amino acids. You don't need to know different R groups.

✓ *Quick check 1*

Hint

Each amino acid has a carboxyl group, an amine group and one of 20 different R groups. R groups and their position in the protein chain determine the functioning of the final protein. They determine the bonds that shape the protein, e.g. enzyme shape determines its active site.

Examiner tip

A polypeptide is a polymer chain of amino acids joined by peptide bonds.

Hint

Hydrolysis involves breaking the bond using water.

Hint

An α helix is a coiled chain and a β pleated sheet is the folded form.

✓ *Quick check 2*

Proteins are macromolecules of **amino acids**. Haemoglobin is the red protein pigment packed into red blood cells.

Key definition

Macromolecules are very large molecules. Triglycerides are a good example of macromolecules. They are made up of only four molecules – a glycerol and three very long fatty acid chains. Many of the other biological molecules such as proteins and polysaccharides are polymers. They are large molecules but made up of many similar smaller molecules (monomers).

Amino acids

Twenty different amino acids are used to make proteins. Amino acids form polypeptides when they join together by **condensation** reactions forming a chain. Each reaction loses a molecule of water and results in a **peptide** bond.

The structure of an amino acid

Condensation and hydrolysis. Two amino acids can be joined to make a dipeptide. The dipeptide can be split back into two amino acids.

Chemically adding water **hydrolyses** the peptide bond.

Proteins and polypeptides

There are four levels of organisation in proteins.

The **primary structure** describes number and order of amino acids in the polypeptide chain. Some proteins have large numbers of amino acids. Haemoglobin has four polypeptide chains each with more than 100 amino acids.

Folding or coiling of a polypeptide chain into alpha helix or beta sheet is the **secondary structure**. Weak hydrogen bonds form between different parts of the chain to stabilise it.

Further folding gives a specific three-dimensional shape. This is its **tertiary structure** which is very specific to the polypeptide function. The shape is stabilised by different types of bond between R groups, including:

- weak hydrogen bonds
- strong disulfide bonds between amino acids with sulfur R groups

- ionic bonds between amino acids carrying a charged R group
- hydrophobic interactions between non-polar R groups.

When proteins are heated to high temperatures or subjected to a change of pH, some of these bonds break causing the protein shape to change and **denature**.

Some proteins have more than one polypeptide chain. These have a **quaternary structure**. Haemoglobin has a quaternary structure with four polypeptide chains held together by hydrogen or ionic bonds. Collagen is a quaternary protein with three polypeptide chains. Antibodies have four polypeptide chains.

Fibrous proteins, like collagen, have polypeptide chains joined to form long fibres held together tightly. Proteins like these are strong and insoluble. They have a structural function, e.g. keratin in hair.

Globular proteins are globular in shape and are soluble with hydrophilic R groups on the outside. They have biochemical functions, e.g. enzymes or antibodies.

The role of haemoglobin

Haemoglobin is a globular protein of four polypeptide chains, each with haem, a non-protein **prosthetic group**, attached. There are two alpha and two beta globin chains.

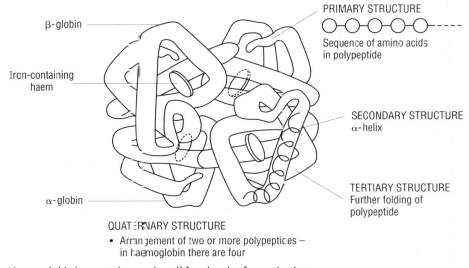

Haemoglobin is a protein showing all four levels of organisation

When oxygen combines with haem the haemoglobin molecule shape will change slightly, making it easier for the next haem group to pick up oxygen.

Small changes in the primary structure of haemoglobin lead to several disorders. Sickle cell anaemia is where the two beta chains contain a different amino acid. When oxygen is short this causes a change in shape and the red cells become sickle shaped. Thalassaemia is where the beta chains are shorter than normal leading to less oxygen transport.

Glycolation of haemoglobin may occur when glucose levels are too high, such as in diabetes. This also leads to oxygen shortage.

✓ *Quick check 3*

Examiner tip

However, you must remember peptide bonds are only broken by hydrolysis.

✓ *Quick check 4*

Examiner tip

A prosthetic group is a non-protein group. In the middle of haem is iron which combines loosely with one oxygen molecule. As there are four chains with haem, one haemoglobin molecule will carry four oxygen molecules. The oxygen binds reversibly.

✓ *Quick check 5*

QUICK CHECK QUESTIONS

1 Draw and annotate a diagram showing how a dipeptide is formed.

2 Describe the difference between primary, secondary and tertiary structures of a polypeptide.

3 Explain what happens when a molecule is denatured.

4 Describe the difference between the quaternary structure of a fibrous and a globular protein.

5 What is the role of haemoglobin?

Water

Key words

- polar molecule
- solvent
- plasma
- tissue fluid
- lymph
- serum
- electrolytes

✓ Quick check 1

Hint

A polar molecule has areas of negative and positive charges. A polar molecule has an uneven distribution of charge such that some regions are slightly (δ) positive and others slightly (δ) negative. The letter δ is used to symbolise 'small' or 'relatively'.

✓ Quick check 2

Examiner tip

Water is called the universal solvent because it dissolves many different solutes. It is the ideal component of a transport medium.

Hint

Serum is the same as blood plasma but fibrinogen, the blood clotting protein, has been removed.

✓ Quick check 3 and 4

Water is an important molecule for life as most biological reactions occur in water. Your body is 60–70% water. It is a **polar molecule** with unique properties.

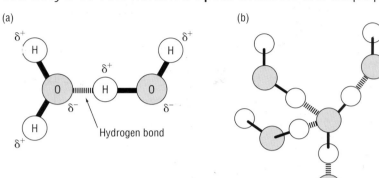

(a) A water molecule showing the uneven distribution of charge; (b) a cluster of water molecules with hydrogen bonds between them

Water is attracted to other water molecules by hydrogen bonds. δ positive hydrogen atoms are attracted to δ oxygen atoms. It is also attracted to other polar molecules. Water is a good **solvent** for polar or charged molecules or ions as it surrounds (hydrates) them making them dissolve.

Properties of water important to life

- Is a good solvent, which allows it to act as a transport medium for polar or charged molecules.
- A high latent heat of vaporisation makes cells high in water very thermostable and allows cooling for thermoregulation.
- High cohesion so molecules 'stick' together. This makes it ideal for support, e.g. amniotic fluid and in plant transpiration and turgidity.
- Has lubricant properties because of its cohesive and adhesive properties, e.g. mucus and pleural fluid.

The differences between plasma, tissue fluid and lymph

Plasma	Tissue fluid	Lymph
Straw-coloured liquid of the blood	The fluid around the cells in tissues. Formed by fluid forced through capillary walls at arterial end	Tissue fluid that doesn't return to capillaries drains into lymphatic vessels
Water-based	Water-based	Water-based
Contains blood proteins, fibrinogen, albumin and antibodies	No large blood proteins pass through capillary walls so these are absent	No large blood proteins
	None of the large blood proteins diffuse through capillary wall	None of the large blood proteins diffuse through
Contains dissolved substances, e.g. ions, glucose, amino acids, oxygen, and waste carbon dioxide and urea	Contains dissolved substances as in the blood. The exact composition depends on the part of the body	Dissolved substances have been exchanged for cell waste such as carbon dioxide and urea from liver cells
Carries hormones to their target organs, e.g. insulin	Carries hormones to cells	Returns used hormones to blood
Transfers heat around body	Transfers heat to tissues	Returns to the blood in a vein in the neck

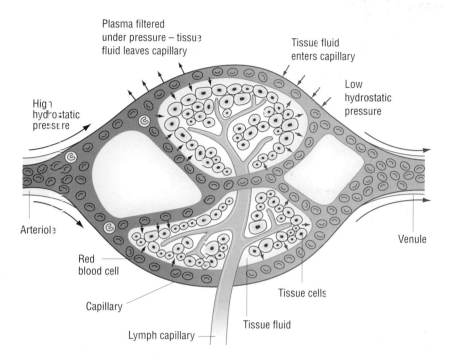

High hydrostatic pressure

Plasma filtered under pressure – tissue fluid leaves capillary

Tissue fluid enters capillary

Low hydrostatic pressure

Arteriole

Venule

Red blood cell

Capillary

Tissue cells

Lymph capillary

Tissue fluid

The relationship between plasma, tissue fluid and lymph

Importance of electrolytes

Water potential balance is affected by dissolved solutes but is mostly maintained by **electrolytes** in the plasma and cells.

The concentrations of electrolytes are vitally important and so are kept at a very steady level. These concentrations are monitored in managing many conditions, e.g. diabetes and kidney disease.

Key definition

Cardiac arrest is when the heart stops beating, possibly caused by a blockage in the blood supply to the heart muscle. Heart failure is the gradual weakening of the heart leading to lung congestion, oedema and liver congestion, amongst other problems.

Testing for electrolytes

Potentiometry measures the voltage across an ion-selective electrode. This is compared to a reference electrode, and so the concentration of the ion is deduced. A sample of blood, plasma or urine is placed in the potentiometer and the readings are delivered quickly and automatically.

QUICK CHECK QUESTIONS

1   Why is water described as a polar molecule?

2   Describe the importance of water in living organisms.

3   In which blood vessel would you expect to find high levels of urea?

4   Explain why fibrinogen and albumin are not present in tissue fluid.

5   How is the water potential balance in the blood maintained?

Water potential and diffusion

Key words

- diffusion
- facilitated diffusion
- osmosis
- water potential
- active transport
- endocytosis
- exocytosis

To ensure cells obtain what they need and waste products are removed, cells exchange substances across their membranes.

A red blood cell is an example of a cell exchanging molecules between the cytoplasm and the surroundings (plasma) using:

- diffusion, including facilitated diffusion
- osmosis
- active transport
- bulk transport.

Diffusion

Diffusion is a passive process requiring no ATP. The molecules move due to random motion down a concentration gradient. Oxygen and carbon dioxide will diffuse directly through the red blood cell phospholipid bilayer as they are small, uncharged molecules.

Facilitated diffusion

Charged molecules (charged ions, such as Na^+ or Cl^-) or water-soluble ones cannot move through the membrane. These molecules diffuse across using a membrane protein.

- Channel proteins: these trans-membrane proteins form pores that allow certain molecules to diffuse across.
- Carrier proteins: these change shape to assist some molecules across the membrane.

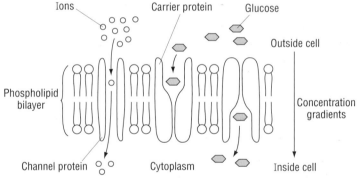

Facilitated diffusion through channel and carrier proteins

✔ *Quick check 1 and 2*

These types of diffusion are called **facilitated diffusion** as they are passive processes with no energy involved. A concentration gradient is needed.

Osmosis

Osmosis is a type of diffusion involving water molecules moving through a partially permeable membrane. Water crosses a membrane by slowly diffusing across the phospholipids. However, cells can contain channel proteins called water pores (aquaporins) in their membranes. This allows water to move across up to 1000 times more rapidly.

Water potential ψ is the term used to explain osmosis.

Key definition

The amount of free energy in the water in a solution is called the water potential. It describes the tendency of water to move. Water flows from high water potential, such as a low concentration of solutes or pure water, to low water potential. A solution with many solutes has a low water potential.

Term	Description of the solution
Isotonic	The same water potential as the cell cytoplasm
Hypertonic	Lower water potential than cytoplasm
Hypotonic	Higher water potential than cytoplasm

✔ *Quick check 1*

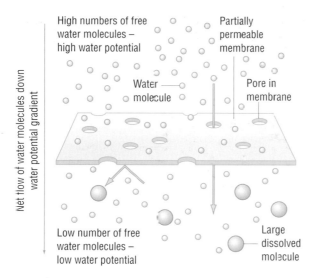

Osmosis

The water potential of the blood

Water potential in the plasma is affected by electrolytes, and by the concentrations of glucose and plasma proteins. A high concentration of these substances lowers the water potential (more negative). This draws water *into* the plasma by osmosis. Low concentrations of these substances raise the water potential (less negative) and water *leaves* the plasma into the surrounding tissues.

Active transport

✓*Quick check 1 and 2*

Active transport is when a molecule is moved across a cell membrane *against* the concentration gradient (from an area of low to one of high concentration). This requires energy from ATP and a specific carrier protein. Many cells, e.g. red blood cells, have a protein called a sodium–potassium pump in their membrane. This pump uses ATP to move sodium ions out of the cell and potassium ions into the cell.

Bulk transport: endocytosis and exocytosis

Larger molecules are moved across membranes by bulk transport. Vesicles take molecules into cells – **endocytosis** – and vesicles within the cell release molecules out of cells – **exocytosis**. Neutrophils use a form of endocytosis called phagocytosis to take in pathogens. The breakdown products are lost by exocytosis.

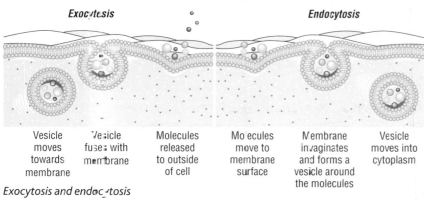

Exocytosis			Endocytosis		
Vesicle moves towards membrane	Vesicle fuses with membrane	Molecules released to outside of cell	Molecules move to membrane surface	Membrane invaginates and forms a vesicle around the molecules	Vesicle moves into cytoplasm

Exocytosis and endocytosis

QUICK CHECK QUESTIONS

1 Describe the terms: diffusion, facilitated diffusion, osmosis and active transport.
2 Give one similarity and two differences between active transport and facilitated diffusion.
3 Explain why red blood cells will burst (haemolysis) when placed in pure water.
4 What is meant by bulk transport and why is it important in living cells?

Carbohydrates

Carbohydrates are molecules containing only atoms of carbon, hydrogen and oxygen. They act as energy stores and energy sources in animals and plants.

Glucose ($C_6H_{12}O_6$) is a **monosaccharide** or simple sugar. Glucose can be built up by condensation reactions into large molecules (**polysaccharides**). Glucose is the sugar found in human blood.

Examiner tip

Learn the glucose structure in the diagram.

Key definition

A respiratory substrate is a substrate molecule that is used during respiration for energy release. Glucose is the most common respiratory substrate, but cells can use other substrates such as fatty acids and glycerols, or even amino acids if necessary, because the body is starved of other molecules.

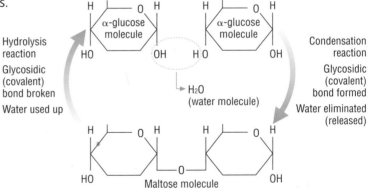

The structure of alpha glucose

Hint

Glucose is the main **respiratory substrate** used to make ATP. It is soluble so is transported easily in blood plasma to the cells.

Disaccharides

A condensation reaction between two monosaccharides forms a **disaccharide**. The bond formed is a **glycosidic bond**, which is covalent. See the diagram below.

- Maltose is two glucose molecules joined by a glycosidic bond.

- Sucrose is a glucose and fructose molecule joined by a glycosidic bond.

Water is needed to hydrolyse (break down) glycosidic bonds and release the monosaccharides.

✔ *Quick check 1*

✔ *Quick check 2*

Hydrolysis reaction
Glycosidic (covalent) bond broken
Water used up

α-glucose molecule

α-glucose molecule

Condensation reaction
Glycosidic (covalent) bond formed
Water eliminated (released)

H_2O (water molecule)

Maltose molecule

Maltose is a disaccharide sugar. It is sweet and soluble.

Making and breaking bonds

Hint

Glycogen is an ideal storage molecule.

- It forms long chains with many branches making glucose release rapid (bonds are broken from the ends of each branch).
- It is compact so stores lots of glucose in a small space.
- It is insoluble so the water potential of the cell is unaffected.

✔ *Quick check 3*

Polysaccharides

Polysaccharides are macromolecules formed by joining many monosaccharides by condensation reactions. Glycosidic bonds join each molecule into a long chain, or polymer.

Glycogen is the polysaccharide stored in human cells such as muscle and liver cells.

Hint

Measuring blood glucose: see page 81.

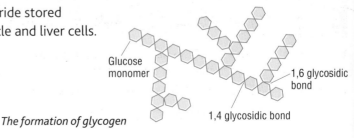

Glucose monomer

1,6 glycosidic bond

1,4 glycosidic bond

The formation of glycogen

Lipids

Triglycerides are molecules, including fats and oils. They consist of glycerol joined to three fatty acid molecules. An ester bond joins them formed by condensation reactions. Fatty acids are macromolecules consisting of long hydrocarbon chains with a carboxyl group attached.

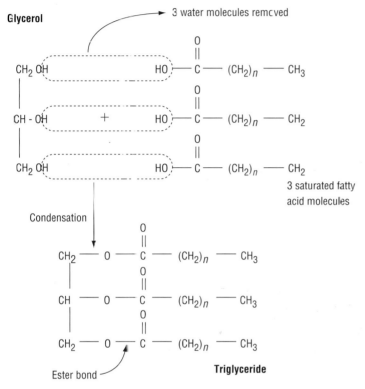

Quick check 4

A triglyceride is formed when three fatty acids form ester bonds with glycerol. This is another example of a condensation reaction

$n = 12$ to 20

Examiner tip

Saturated fatty acids have all single carbon-to-carbon bonds in the hydrocarbon chain. If there is a double bond between carbons there are fewer hydrogen atoms and the chain bends. This is an **unsaturated** fatty acid. A polyunsaturated fatty acid contains more than one double bond. Double bonds result in kinky tails in the fatty acid chain.

Hint

Unsaturated fatty acids:
- lead to fewer fatty deposits in the artery walls if they form the bulk of dietary fat intake
- have lower melting point and remain a liquid at lower temperatures.

✓*Quick check 5*

Role of lipids

- Main energy store in human adipose tissue. Contain more energy per gram than carbohydrates. Insoluble in water.
- Store of fat and fat-soluble vitamins A and D in the liver cells.
- Under the skin, act as a heat insulator.
- Act as an insulator around nerve cell axons, preventing ions leaking.
- Protect delicate organs, e.g. kidneys.
- Cholesterol is a sterol important in cell membranes. Steroids such as the sex hormones are synthesised from cholesterol. Cholesterol is not a triglyceride but is often included with lipids as it is lipid soluble.

QUICK CHECK QUESTIONS

1   Describe the structure of monosaccharides, e.g. glucose, and their importance in human blood plasma.

2   State how a glycosidic bond forms and how it is broken.

3   What is the importance of glycogen in humans?

4   Describe the structure of a triglyceride molecule.

5   Describe the difference between a saturated and an unsaturated fatty acid.

Enzymes and blood loss

Key words

- platelets
- thromboplastin
- prothrombin
- thrombin
- fibrinogen
- fibrin
- activation energy
- complementary
- enzyme–substrate complex
- enzyme concentration

✓ *Quick check 1 and 2*

✓ *Quick check 3*

Hint

Calcium ions are needed for all of these processes. Calcium is an enzyme cofactor for thromboplastin and thrombin. The removal of calcium ions prevents blood clotting during storage.

Examiner tip

Enzyme inhibitors slow the rate of enzyme-controlled reactions by either blocking the active site or causing it to change shape. In both cases ESCs cannot form. Some drugs, for example aspirin, act by inhibiting enzymes.

Blood clotting

When a blood vessel is damaged a chain of reactions occurs to create a clot. The clot prevents excess blood loss and entry of pathogens.

- Damaged tissues expose collagen fibres to the air.
- Blood **platelets** become trapped in the fibres, making other platelets sticky and a plug forms.
- The tissues and damaged platelets release the enzyme **thromboplastin**.
- Thromboplastin converts the blood protein **prothrombin** into an active enzyme, **thrombin**.
- Thrombin converts the blood protein **fibrinogen** into long insoluble **fibrin** fibres.
- The fibres form a mesh that traps blood cells to form a blood clot. This dries into a scab.

All enzymes catalyse specific reactions. The reaction occurs faster than normal at body temperature. Enzymes do this by lowering the **activation energy** needed.

Each type of enzyme has a specific tertiary shape and an active site. A specific substrate fits into the specific active site because it has a **complementary** shape. This forms an **enzyme–substrate complex** (ESC). This allows the enzyme to exert forces and lower the energy required for the reaction.

Key definition

Activation energy is the amount of energy needed to overcome the energy barrier in a chemical reaction. This barrier is usually quite large and so reactions will only occur very slowly over time or at very high temperatures. Enzymes reduce the activation energy when they form an enzyme–substrate complex. This allows reactions to occur very quickly and at lower temperatures.

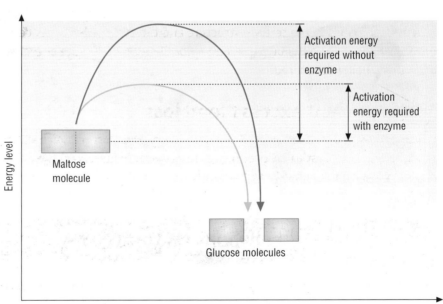

Adding the enzyme maltose reduces the amount of activation energy required for the reaction to take place

How enzymes lower activation energy

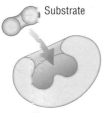  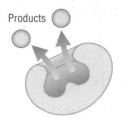

Substrate

Products

Enzyme

Enzyme–substrate complex

Lock and key mechanism of enzyme action. In this case, the enzyme splits (hydrolyses) the substrate molecule into two smaller products. it is worth noting that given the right circumstances, the enzyme can work in reverse.

Mechanism of enzyme action

Factors affecting enzyme action

- **Enzyme concentration** affects the rate of an enzyme-controlled reaction. With more enzyme molecules, the rate of reaction is faster because more active sites are available. More ESCs can form per second and so more product is made.
- Low **substrate concentration** gives a slow rate. As more substrate is added the rate increases. At very high levels the rate will be limited by the concentration of enzyme. All the enzyme active sites will be occupied. The reaction is only as fast as the enzymes can catalyse the reaction and free the active site.
  A disorder with low blood fibrinogen levels results in slow clotting time with small amounts of fibrin formed per second.
- At low **temperature** the rate is slow because there is little kinetic energy available. As the temperature rises there is more kinetic energy and the enzyme and substrate molecules collide more frequently and form more ESCs. Above the optimum temperature the enzyme tertiary shape begins to distort and at high temperatures the enzyme denatures and the active site is no longer complementary to the substrate.
  Blood is stored at $4\,°C$ to prevent enzyme activity and keep the blood stable. Freezing would rupture the blood cell membranes as ice forms.
- **pH** is a measure of acidity or alkalinity. It affects enzyme activity since a change in pH affects ionic bonds in the tertiary structure and the enzyme becomes denatured. In blood, changes in pH must be prevented by buffering to avoid the enzymes and blood proteins being denatured.

First aid to prevent excess blood loss

If excessive bleeding occurs, such as after a severe injury, the clotting mechanism is not enough to stop the flow. First aid procedures are needed to help reduce the flow. You must also seek medical help quickly and urgently.

QUICK CHECK QUESTIONS

1 Name the three main blood proteins found in the plasma.

2 Draw a flow diagram to illustrate the steps involved in the blood clotting process.

3 Explain why the enzyme thromboplastin forms enzyme–substrate complexes with prothrombin and produces thrombin, but will not form a complex with other proteins.

4 Explain why is it important to keep the pH and temperature constant when storing blood.

5 Why will the clotting mechanism not stem blood loss after a severe injury?

UNIT 1

Blood transfusions

Blood transfusions are frequently vital in saving lives. It is important that the blood donated is stored correctly to prevent clotting or other enzyme or osmotic changes.

Key definition

Blood transfusions are life-saving events in many patients. However, the donated blood must be carefully matched to the patient's blood to avoid the patient's immune response attacking and destroying the new transfused cells.

The donated blood must also be carefully screened and tested to avoid transmitting any harmful infections such as HIV and hepatitis B.

✓ Quick check 1

Storing blood

The enzymes in blood are affected by several different factors such as temperature and pH. See page 17.

Cofactors are substances needed for enzyme reactions. Calcium ions are an example of a cofactor needed by several of the enzymes in blood clotting.

✓ Quick check 2

Removing these calcium ions prevents clotting enzymes from working.

Conditions needed to store blood:
- low temperature (4 °C)
- pH **buffering**
- removal of calcium ions using a chemical such as sodium citrate
- storage in sterile plastic bags consisting of single units of blood for easy use
- mannitol and sodium chloride (SAGM) to maintain a water potential balance with packed red cells.

✓ Quick check 3

Hint

Citrated blood is blood treated with sodium citrate to remove calcium ions.

Screening blood

Donated blood is screened to avoid passing on infections. The donor is first asked health-related questions. A sample of the donated blood is tested for infections including HIV and hepatitis C. It is also tested to find the blood group.

✓ Quick check 4

Testing for infections involves testing for antibodies and testing for viral nucleic acid. HIV nucleic acid is present even before the antibodies are present.

Examiner tip

Transmission of infections from infected blood can also be prevented by producing clotting factors using recombinant DNA technology. (This is an A2 topic, and will be covered more fully on page 130.)

Treating blood

Blood products are treated to destroy viruses. Increasing concentrations of cold ethanol is one method used. Heating blood products can also be used if the product is heat stable.

Blood groups

An individual's **blood group** is determined genetically. The ABO system is one of several blood group systems. These blood groups are distributed differently around the world. Blood group O is the most common group in the UK at 63%, but in South America it is nearly 100%. Group AB is the rarest with only 3% in the UK, but rises to 20% in Eastern European countries. (This is covered more fully on page 64.)

Types of blood products

Some of the donated blood is processed into different forms.

Types of stored products	Uses	Description
Whole blood	For severe blood loss. Not much used	Contains everything normally in blood Maintains blood volume in recipient
Leuco-depleted	For patients needing frequent transfusions, e.g. patients with aplastic anaemia need regular transfusions	Most of the leucocytes removed Maintains blood volume in recipient
Packed red cells	For anaemic people or to replace blood loss after surgery or childbirth	Red blood cells separated off and stored Can be added without increasing blood volume
Plasma	During cardiac surgery, during childbirth and to replace clotting factors	All blood cells removed Can be frozen Maintains blood volume in recipient
Platelets	For patients with bone marrow failure and for patients after transplants and chemotherapy	Platelets separated off and stored
Clotting factors	For patients with clotting disorders such as haemophilia	Blood plasma is processed to provide clotting factors

✔ *Quick check 5 and 6*

QUICK CHECK QUESTIONS

1 What affect could osmotic changes have on stored blood?

2 Explain the importance of cofactors such as calcium ions.

3 Describe the conditions needed to store blood in perfect condition.

4 Describe the different ways in which blood is screened, tested and treated to avoid transmission of diseases such as HIV.

5 Name the different types of blood products made from donated blood.

6 Why is using whole blood for transfusion a rare occurrence nowadays?

The heart

Key words

- cardiac cycle
- myogenic
- sinoatrial node
- atrioventricular node
- ECG
- Purkyne tissue
- cardiac output
- stroke volume
- pulse rate

✓ *Quick check 1*

Structure of the human heart

The heart pumps blood around the body. The heart is two pumps side by side:
- left side receives blood from the lungs and pumps it to the body
- right side receives blood from the body and pumps it to the lungs.

Blood enters the two atria which contract and force blood into the two ventricles. The ventricles then contract and blood is pumped out into the arteries. The right ventricle pumps blood into the pulmonary artery and up to the lungs (the pulmonary circulation). The left ventricle pumps blood into the aorta and round the body (the systemic circulation). The atria contract together and the ventricles contract together throughout life.

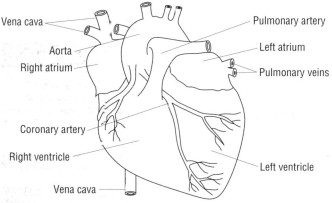

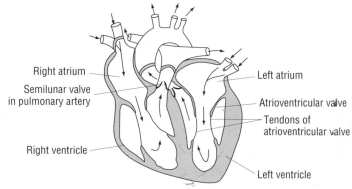

External view of the human heart showing the four chambers, main blood vessels, and coronary arteries that supply blood to heart muscle

Vertical section through the heart to show the internal structure. The left and right atria have much thinner walls than the ventricles because the atria contract to push blood into the ventricles at low pressure, to ensure that they are full

Examiner tip

Make sure you learn the heart structure from the diagrams. The flow of blood through the heart is shown as arrows on the internal structure diagram.

✓ *Quick check 2*

The cardiac cycle

The **cardiac cycle** is one complete cycle of events beginning with diastole, followed by atrial systole, and finally ventricular systole. The cycle is repeated about 72 times per minute.

Name	Event	Outcome	Valves	Pressure
Diastole	Heart muscle is relaxed	The atria fill with blood, opening the atrioventricular (AV) valves, and blood enters ventricles	Semilunar valves close AV valves open	Low but slowly rises as blood enters Low in arteries (diastolic pressure)
Atrial systole	Atrial muscle contracts	Blood pushed into ventricles Atria are emptied	AV valves pushed fully open	Rises in both atria and ventricles as blood enters
Ventricular systole	Ventricular muscle contracts	Blood forced into the arteries (right ventricle into pulmonary artery and left ventricle into aorta)	AV valves snap shut Heart tendons hold valves in place Semilunar valves pushed open as blood enters arteries Valves pushed open when ventricular pressure is greater than arterial pressure	Atrial pressure falls Ventricular pressure rises steeply as muscle contracts Continues to rise for 0.1 second but falls as empties Artery pressure is high (systolic pressure)

Hint

Closure of the valves prevents backflow. The AV valves snapping shut gives first part of heart sound (lub). Closure of semilunar valves gives second (dub) part of heart sound.

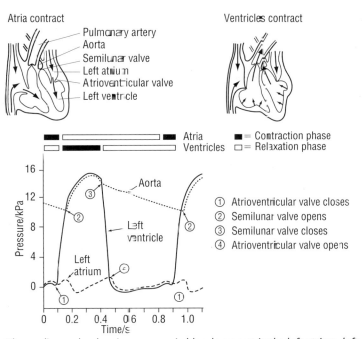

Atria contract

- Pulmonary artery
- Aorta
- Semilunar valve
- Left atrium
- Atrioventricular valve
- Left ventricle

Ventricles contract

■ = Contraction phase
□ = Relaxation phase

Atria
Ventricles

① Atrioventricular valve closes
② Semilunar valve opens
③ Semilunar valve closes
④ Atrioventricular valve opens

The cardiac cycle, showing changes in blood pressure in the left atrium, left ventricle and aorta during one heartbeat and the beginning of the next

The stages of the cardiac cycle

Ventricular systole | Diastole | Atrial systole

An ECG

- The heart wall is made of special muscle – cardiac muscle – it is **myogenic** and needs no stimulation to contract.
- The **sinoatrial node** in the right atrial wall generates an electrical impulse that travels across the atria and triggers atrial systole.
- The **atrioventricular node** detects this and, after a short delay, relays it onto the ventricles via **Purkyne fibres** in the septum. The impulse then spreads across the ventricles to trigger ventricular systole. Systole is always followed by diastole – the period when the heart muscle is recovering and relaxed.
- An **ECG** (electrocardiogram) monitors the electrical impulse as a trace. Differences in the trace indicate heart problems.

Cardiac output is the volume of blood pumped out of the left ventricle per minute.

Cardiac output = **stroke volume** (the volume pumped with each beat) × heart rate

Increase in cardiac output will result during strenuous exercise as the body muscles contract more and require more oxygen for aerobic respiration.

A normal ECG trace

Heart rate can be measured by taking the **pulse rate**. The pulse is the expansion and elastic recoil of the artery wall as the blood is pumped out of the heart under pressure. It is measured at the wrist or in the neck.

QUICK CHECK QUESTIONS

1 Why is the heart described as a double pump?
2 Describe the complete passage of a single blood cell through the body.
3 Use the graph to find the time taken for one heartbeat in the cardiac cycle and calculate the heart rate.

4 Explain what information can be gained from an ECG.
5 Explain heart rate and describe how changes in rate will result in changes in cardiac output.

Circulatory system

Key words

- mass transport
- arteries
- veins
- capillaries
- blood pressure
- sphygmomanometer

Examiner tip

With mass flow everything moves in the same direction. In human blood the plasma and all the blood cells all move together.

Examiner tip

The system is **closed** – blood always moves within blood vessels.

You will remember it is a **double** system from your study of the heart. There is one circuit between the lungs and the heart (the pulmonary circuit) and one between the body and the heart (the systemic circuit). The blood passes through the heart twice during one full circuit around the body.

✓ *Quick check 1*

Mass transport

The human circulatory system acts as a transport medium by mass flow.

A closed and double circulatory system

Transport in humans needs a circulatory system because diffusion alone is not adequate to deliver enough oxygen and nutrients to all respiring cells. The diffusion path is too long to allow enough substances such as glucose to move quickly enough.

The circulatory system is described as a closed double system as seen in the diagram.

Deoxygenated blood returns from the body to the heart and then on to the lungs. Oxygenated blood from the lungs returns to the heart and then on to the body.

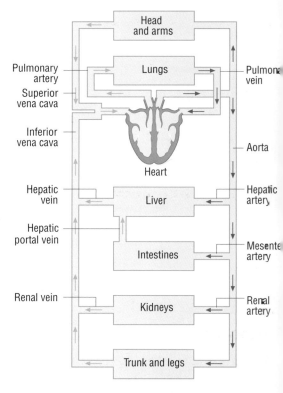

Some of the main blood vessels in the blood circulatory system

Structure and function of arteries, veins and capillaries

Blood vessels	Structure	Function	Blood flow
Arteries	Thick wall of smooth muscle and elastic fibres	Keeps pressure high; blood is distributed throughout circuit Artery wall stretches and recoils	Leaves the heart Rapid flow with high pressure which pulses
Veins	Thin wall with a small amount of smooth muscle and elastic tissue Semilunar valves along the length	Returns blood to the heart Valves prevent backflow and keep flow moving on	To the heart Flow is slow with low pressure No pulse
Capillaries	No elastic or muscle tissue Wall of single layer of cells of squamous epithelium Lumen only 7–10 μm wide	Form a network within tissues and organs Allow exchange of metabolites between the blood and surrounding tissues	Connect arterioles and venules Slow rate of flow Low pressure
Arterioles	Thin wall mainly of smooth muscle and some elastic fibre	Carry blood between arteries and capillaries Regulate flow and distribution	Smooth muscle contracts to narrow (constrict) the lumen and decrease the flow to certain tissues OR relaxes to widen (dilate) the lumen to increase the blood flow
Venules	Very thin wall of muscle and elastic fibres	Carry blood from capillaries back to the veins	To veins from capillaries

✓ *Quick check 2 + 3*

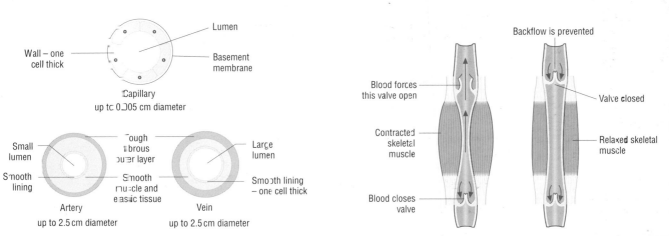

Sections through the vessels. The veins contain valves to make sure the blood flows in the right direction

Relationship between different blood vessels

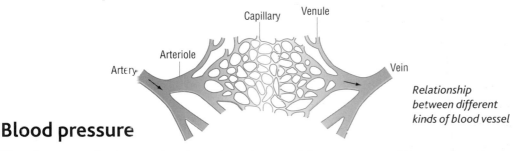

Relationship between different kinds of blood vessel

Blood pressure

Blood pressure in the arteries rises when the ventricles contract and drops when they relax. Elastic recoil of the artery walls gradually smoothes out the pressure when the blood reaches the arterioles. Blood pressure drops as the blood moves through the capillaries. In the veins, blood returns to the heart under very low pressure.

Sphygmomanometer

Blood pressure is measured using a **sphygmomanometer**. The units used are kilopascals (kPa) using an electronic instrument, but most doctors prefer to use a manometer calibrated in millimetres of mercury (mmHg). Blood pressure is taken in the arm using an inflating cuff attached to the instrument and a stethoscope. The readings are given as two values, e.g. 120/80 mmHg. The top value is the systolic and the bottom value the diastolic reading.

High blood pressure is called **hypertension**.

Key definition

Doctors use a sphygmomanometer to measure blood pressure. It is an instrument using a column of mercury to measure the pressure and is considered to be very accurate. A stethoscope is used to listen for the sounds of the blood flowing within the artery to indicate the systolic and diastolic pressure. A reading of pressure as mmHg is obtained. Electronic blood pressure monitors are available and measure pressure in kPa. However the readings are not as accurate for a single reading.

Examiner tip

Check tissue fluid and lymph covered on page 10.

Examiner tip

There are three layers in the walls of arteries and veins: the Tunica externa, the Tunica media and the Tunica intima. The composition of these is described on the next page.

✓ *Quick check 4*

Examiner tip

The slow rate of blood flow through the capillary network ensures exchange of materials can occur across the large surface area.

✓ *Quick check 3*

✓ *Quick check 5*

Examiner tip

Learn all the steps used to measure blood pressure.

QUICK CHECK QUESTIONS

1 State the advantages of a double circulation and a closed system.
2 How do arterioles regulate blood flow to tissues? Give an example in the skin.
3 Explain the features that allow nutrients and gases to diffuse across the capillaries efficiently.

4 How does elastic recoil in the artery wall smooth out the pulsing of the blood?
5 What is hypertension and what blood pressure values would suggest hypertension?

The lungs

Cells, tissues and organs

In animals, cells are specialised to carry out a specific function and are grouped together into a **tissue**.

Tissues are grouped into **organs**. Humans have many different organs, e.g. heart, lung and kidney.

Tissue types

There are several important tissues in the lungs.

- **Squamous epithelial** tissue is flattened and reduces the diffusion distance for efficient gas exchange. It forms the single layer of the alveoli walls.
- **Ciliated epithelial** tissue contains ciliated cells; these have many cilia which beat in rhythm to move particles trapped in mucus out of the lung. It also contains goblet cells which produce a glycoprotein and secrete it in mucus, which traps dirt and bacteria in the air breathed in.
- Ciliated cells and goblet cells form the layer lining the trachea and bronchi. Ciliated cells line the bronchioles.

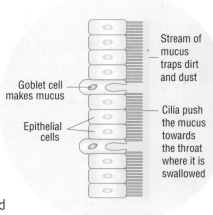

A mucous membrane from the trachea

Examiner tip

A tissue is a group of similar specialised cells carrying out specific functions, e.g. ciliated epithelium. Squamous epithelial tissue is the single layer of cells in the lung alveoli.

Examiner tip

An organ is a group of different tissues forming a distinct structure carrying out a specific function.

✓ *Quick check 1*

Lung structure

Air breathed in moves down the trachea into two bronchi. It then moves into the many branches of the bronchioles and on into the blind ends, the air sacs. These are a cluster of alveoli.

Follow the air passage in the diagram.

Surface area to volume ratio

The ratio between surface area and volume determines the rate of diffusion. A high ratio gives faster diffusion. Large organisms have a small ratio and a long diffusion distance. However, organs such as the lung provide the additional surface area (cells in contact with the environment) needed to allow rapid gas exchange.

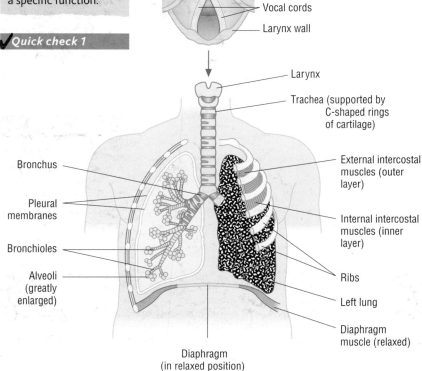

The respiratory system

✓ *Quick check 2*

Gas exchange surface of lungs

Alveoli are the gas exchange surface. They have a single layer of squamous cells for rapid diffusion and elastic fibres for expansion and recoil during breathing.

A good **gas exchange** site has:
- a large surface area
- a thin layer giving a short diffusion distance
- a steep diffusion gradient maintained by ventilation and a good blood supply.

Measuring lung capacity

A **spirometer** measures **lung capacity**. The actual volume of air breathed in depends on the size of your lungs, your health and your activity levels. Asthmatics may need to check the rate they breathe out with a peak flow meter.

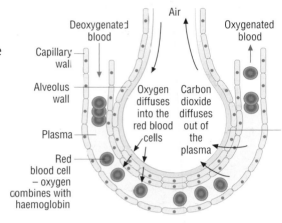

Gas exchange in the alveolus

✔ *Quick check 3 + 4*

✔ *Quick check 5*

Lung capacities

Key definition

A spirometer is used to measure some lung volumes. It is frequently a volume of air trapped under water within a floating chamber. As air is breathed out and in the chamber will rise and fall. A pen attachment records the movement of the chamber as a spirometer trace. Since carbon dioxide is absorbed the volume of air in the chamber decreases over time and so the trace falls slowly.

Hint

Smoking damages the lung by destroying the cilia lining the airways. Measuring lung capacity gives an indication of the degree of damage.

The diagram shows a trace taken when using a spirometer. Residual volume cannot be measured this way. Lung volumes can be read off the y-axis. A time scale on the x-axis allows breathing rate to be calculated.

Respiratory arrest

This is when a person stops breathing. It may be due to:
- respiratory conditions, e.g. asthma or pneumonia
- an airway blockage
- drug overdose or an accident making the person unconscious.

Examiner tip

Learn the terms **tidal volume**, **vital capacity** and residual volume from the diagram.

Examiner tip

Resuscitation is a first aid procedure used on a person with a pulse but not breathing. There are a series of steps you must learn for adults and for children.

QUICK CHECK QUESTIONS

1. State the definitions for tissues and organs and give two examples of each.

2. Explain why larger organisms have a long diffusion distance.

3. Describe the features of alveoli making them a highly efficient gas exchange site.

4. Describe the features of the lung providing an efficient gas exchange surface.

5. How can a peak flow meter be used to check the condition of an asthmatic?

1 In the UK, scientists working for the National Blood Transfusion Service must have a detailed knowledge of red blood cells and their membrane properties.

(a) Briefly describe the structure of the red blood cell membrane. (3)

(b) The four definitions below refer to ways in which molecules and ions may pass through a red blood cell membrane.

A: the movement of water down a water potential gradient through a partially permeable membrane.

B: the net movement of molecules or ions down a concentration gradient.

C: movement against the concentration gradient, which requires an input of energy from the cell.

D: passive movement across a cell membrane with the aid of transport proteins.

Name the processes **A** to **D**. (4)

(c) The National Blood Transfusion Service has to be able to store blood products for use in emergency situations.

Under normal conditions in the body, red blood cells actively pump potassium ions into and sodium ions out of the cell.

If blood is kept in cold storage, red blood cells lose potassium ions to the external plasma.

Suggest why potassium ions may accumulate in the plasma in cold storage. (2)

(d) Before storage, blood is screened for HIV and hepatitis. A prospective donor will be asked questions before giving blood to see if they are suitable.

(i) As part of the screening process, the donor's blood will be mixed with antigens for HIV and hepatitis. If the donor has come into contact with HIV or hepatitis, their blood will react with the antigens.

Explain why a reaction occurs when blood is mixed with HIV or hepatitis antigens. (2)

(ii) Suggest *two* reasons, other than exposure to HIV or hepatitis, why a prospective donor may not be allowed to give blood. (2)

2 (a) Figure 1 shows the structure of an alveolus.

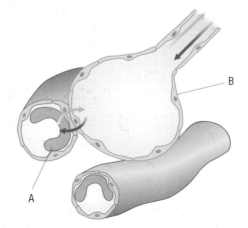

(i) Name cells **A** and **B**. (2)

(ii) Surfactant is found on the inner surface of the alveoli.

Explain the role of surfactant on the inner surface of the alveoli. (2)

(b) Cyclists in the Tour de France spend some time at altitude in the Alps during the competition.
Figure 2 shows two spirometer traces from the same cyclist, taken before and after the longest climb in the competition, Alpe d'Huez.

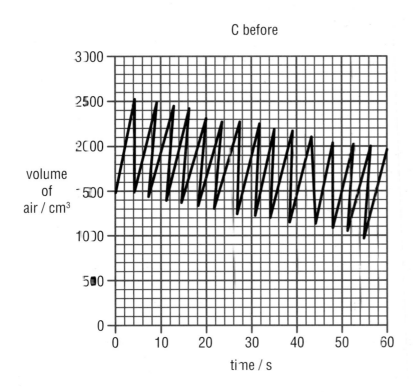

C before

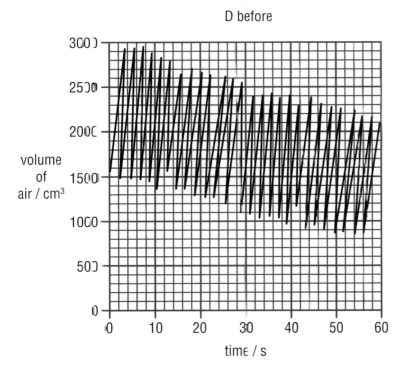

D before

(i) State the differences in breathing shown on the two traces. (1)

(ii) What is the number of breaths per minute recorded on trace **D**? (1)

(iii) State *two* safety precautions which must be taken when using a spirometer. (2)

(iv) Suggest why both traces slant downwards. (2)

3 In the past, some athletes have used a banned technique called blood doping to improve performance. This involves increasing the number of red blood cells just before competition, which increases the oxygen-carrying capacity of the blood. Authorities can test for this by measuring the concentration of haemoglobin in the blood as a method of determining the number of red blood cells present.
Describe a procedure which may be used to determine the number of red blood cells in a human blood sample. (7)
In this question, one mark is available for the quality of use and organisation of scientific terms. (1)

4 Glycogen is an important storage molecule in the body.
Figure 3 shows part of a glycogen molecule.

Complete the following passage about glycogen by using appropriate words from the list to fill the gaps.

**sugar condensation polysaccharide ester glucose
muscle glycosidic hydrolysis water
sucrose fat respiration energy**

Glycogen is a .... made from hundreds of .... molecules.

In order to join these molecules together, ... bonds are formed by a reaction called ....

This type of reaction requires the removal of a molecule of ....

We store glycogen in liver cells and .... cells. These glycogen molecules are a short-term .... store that the body can use when muscle cells are working hard. The glycogen molecules can easily be broken down to their constituent monosaccharides. Each glycogen molecule has lots of ends and its subunits are easily broken off for use in .... (8)

5 Aspirin is a well known painkiller and was introduced in 1899. However, not until recently have other possible uses been suggested for this drug.
- Aspirin and similar drugs work by inhibiting the production of prostaglandins.
- Prostaglandins are local hormones produced from phospholipids.
- Prostaglandins increase blood flow to an infected area, so that there are more blood cells to combat infection.
- Aspirin also inhibits the enzyme cyclo-oxygenase (COX. A2 only); this reduces the amount of prostaglandins produced. Prostaglandins are known to increase blood clotting

(a) Name a white blood cell that combats infection. (1)
(b) How may reducing prostaglandins decrease heart disease caused by blood clots? (5)
(c) Outline the normal process of blood clotting. (4)
(d) Suggest why low doses of aspirin, taken regularly, may be used to treat patients with coronary heart disease (A2 only). (2)

6 (a) Explain what is meant by cardiac output. (2)
(b) Table 1 shows the percentage of total blood volume delivered to various organs during light, moderate and maximum exercise.
Table 1

	% of total blood volume to organs		
Exercise level	Brain	Muscles	Skin
Light	8	47	15
Moderate	4	71	12
Maximum	3	88	2

Describe *and* explain the trends shown by the data in the table. (3)

(c) Table 2 shows measurements from four 21-year-old students from the same university. Three were training for specific sporting events, whilst one did not train.
Table 2

	Measurement		
Sporting event	Heart septum (mm)	Left ventricle wall (mm)	Stroke volume (cm^3)
800 metre run	11.0	12.0	160.0
Swimming	10.9	10.6	181.0
Shot putt	11.0	12.0	110.0
Did not train	9.5	10.5	82.0

(i) Calculate the percentage increase in size of the left ventricle wall between the 800 metre runner and the student who did not train.
Show your working. (2)
(ii) Training increases stroke volume.
Suggest why training increases stroke volume. (2)
(iii) Suggest *two* reasons why it is unfair to make any valid comparison between the four students. (2)

Mitosis as part of the cell cycle

Key words

- interphase
- mitosis
- semi-conservative replication
- hydrogen bonding
- complementary base pairing
- centrioles
- chromatids
- chromatin
- chromosomes
- diploid
- prophase
- centromere
- nuclear envelope
- metaphase
- spindle fibre
- anaphase
- telophase

The cell cycle is divided into **interphase** and **mitosis**. Mitosis is only about 5–10% of the cell cycle. Interphase is subdivided into two G (gap) phases and an S (synthesis) phase.

Key definition

The cell cycle describes the events that take place in a eukaryote cell leading to its replication. It includes interphase (G_1, S and G_2 phases) followed by M (mitosis and cytokinesis).

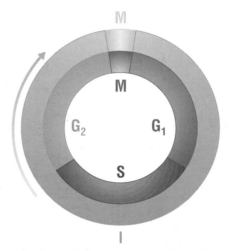

✓ *Quick check 1*

G phases indicate growth of organelle numbers and cell size overall
S indicates synthesis of new DNA (i.e. DNA replication)

M is nuclear division (mitosis) and cytokinesis (cleavage of cytoplasm).
G_1 is the first gap phase where biosynthesis occurs – proteins are synthesised and organelles replicate. S is the synthesis phase where all DNA molecules are replicated. G_2 is the second gap phase where the cell continues to grow and builds up energy reserves needed for mitosis.

Part of a DNA molecule. The two antiparallel strands are held together by hydrogen bonds

Bonding in a DNA molecule

✓*Quick check 2*

✓*Quick check 3 and 4*

✓*Quick check 5*

G_1 (first gap) phase	S (synthesis) phase	G_2 (second gap) phase
Proteins are synthesised	DNA replicates by **semi-conservative replication**:	Cells and organelles reach full size
Organelles replicate	• each molecule unwinds and unzips as the **hydrogen (H) bond**s between **complementary base pairs** break	Synthesis continues
More new cytoplasm is made		The cell builds up energy reserves
G_1 checkpoint	• free nucleotides pair up to the exposed nucleotide bases on the two DNA strands. There is complementary base pairing. A pairs with T by two H bonds. G pairs with C by three H bonds. This is catalysed by DNA polymerase. DNA ligase catalyses the condensation reactions between a sugar group and a phosphate group of adjacent nucleotides	The **centriole** divides into two
Cyclins and cyclin-dependent kinases (CDKs) interact		G_2 checkpoint
Cell cleared to move into S phase		Cyclins and CDKs interact
		Cell cleared to enter mitosis
	• by the end of the S phase each molecule of DNA has been copied into two identical molecules (the cell checks these are identical to the original molecule); these become the **sister chromatids**	

Stages of mitosis

Prophase

- The chromatin (DNA of chromosomes) is condensed and tightly coiled into chromosomes. Because the DNA has replicated during interphase each visible chromosome consists of two identical **sister chromatids** joined by a **centromere**.
- The **nuclear envelope** breaks up and disperses.
- The two centrioles migrate to opposite ends of the cell, making the poles.

Metaphase

- A **spindle** (made from microtubules) forms.
- Chromosomes (pairs of sister chromatids) attach to the equator of the spindle, by their centromeres.

Anaphase

- Spindle threads shorten, causing each centromere to break, separating each chromatid from its sister.
- The separated chromatids (now called chromosomes) are pulled to opposite poles.

Telophase

- Nuclear envelopes form around each group (double set) of chromosomes at either end of the cell.
- The chromosomes uncoil.
- There are now two diploid nuclei, each genetically identical to the other, and to the nucleus of the parent cell.

The cell now undergoes cytokinesis. Microtubules form a 'draw string' just inside the plasma membrane and this pinches inwards and fuses.

(Diagram labels: Interphase, Prophase, Centriole, Metaphase, Early anaphase, Late anaphase, Telophase, Cytokinesis (division of the cytoplasm))

G_1 = Gap phase 1
G_2 = Gap phase 2
S = Synthesis of DNA (replication)
PMAT = Stages of mitosis
C = Cytokinesis

The mitotic cell cycle in a human cell. Mitosis occupies 5–10% of the cell cycle. At the end of cytokinesis, the products of such a mitotic division are two diploid (having two copies of each chromosome) cells, each genetically identical to each other and to the parent cell.

Hint

During interphase the DNA of the nucleus is in the form of **chromatin** – it is diffuse and *not* condensed or tightly wound into **chromosomes**.

Hint

One chromosome consists of one molecule of DNA (and histone proteins). DNA condenses by coiling round these histone proteins.

QUICK CHECK QUESTIONS

1. Suggest why the term 'cell cycle' cannot be applied to prokaryote cells.

2. Suggest why a cell needs a lot of ATP by the end of G_2 phase.

3. Explain why a purine (A or G) always pairs with a pyrimidine (T or C) in DNA.

4. A new piece of DNA has 3000 thymine bases and 2000 guanine bases. How many deoxyribose groups does it have?

5. When does the cell have double the normal amount of DNA: (a) at beginning of S phase, (b) at beginning of G_1 phase, (c) at beginning of G_2 phase?

Apoptosis

Key words

- apoptosis
- programmed cell death
- necrosis
- growth
- development
- repair
- cell addition
- cell deletion
- vesicles
- endocytosis
- phagocytosis

Hint

Necrosis is the disorderly cell death that results from trauma – injury or toxic chemicals.

✔ *Quick check 2*

Key definition

Apoptosis (pronounced: *ape o toe sis*) is also known as **programmed cell death**. It occurs in multicellular organisms. When a cell has undergone about 50 mitotic divisions, it undergoes an orderly and tidy cell death – this is apoptosis.

✔ *Quick check 1*

Apoptosis is essential for normal **growth** and **development** and **repair**.

- During tissue development there is extensive division and **cell addition** by mitosis, followed by **cell deletion** through apoptosis. The apoptosis is tightly regulated by signals from surrounding tissues or signals from within the cells.
- In an embryo the hands and feet are webbed. During development, cells between fingers on the hands and between toes on the feet undergo apoptosis, so that at birth the digits are separated from each other.
- In children between 8 and 14 years of age, 20–30 billion cells per day apoptose. This is outweighed by the number being produced by mitosis, as the children are growing.
- In adults, between 50 and 70 million cells per day apoptose. This should be balanced by the rate of cell production by mitosis. If apoptosis outweighs mitosis, cell loss and degeneration occurs (e.g. loss of heart muscle cells after a myocardial infarction).
- When a broken bone mends, new bone cells are produced – more than are needed – by mitosis. The extra cells are deleted by apoptosis.

Apoptosis can also be triggered by other events such as:

- infection with a virus. Human immunodeficiency virus (HIV) enzymes prime T helper cells for apoptosis, while delaying it long enough for viruses to be replicated in the T cell. Cells infected with other viruses display viral antigens to stimulate T killer cells to initiate apoptosis
- stress, such as when a cell is starved of nutrients or oxygen
- DNA damage
- increased calcium ion concentration in a cell as a result of damage to the cell surface membrane.

The sequence of events during apoptosis of a cell is very quick and is as follows.

1 Enzymes break down the cell cytoskeleton.
2 The cytoplasm becomes dense, with organelles tightly packed.
3 The cell surface membrane changes and small bits, called blebs, form.
4 Chromatin condenses and the nuclear envelope breaks.
5 DNA is broken into fragments by endonuclease enzymes in the cell.

✔ *Quick check 3*

6 The cell breaks up into **vesicles**.
7 The vesicles are ingested, by **endocytosis**, by phagocyte cells (**phagocytosis**). This ensures that cellular debris does not damage any nearby cells. The cellular components may also be reused.
8 There is *no* inflammatory reaction and *no* hydrolytic enzymes are released.

✔ *Quick check 4*

Control of apoptosis

Some genes are involved.

- p53 is a protein made by a tumour suppressor gene, TP53. p53 inhibits mitosis and leads to cells undergoing apoptosis. A mutation of the TP53 gene produces an altered p53 protein and this can lead to cancer – an uncontrolled cell division – because cells fail to apoptose.
- Other genes make proteins that bind to apoptosis-inhibiting factors, triggering apoptosis.

Signals from inside the cell may be involved.

- Nitric oxide makes the inner mitochondrial membrane more permeable to hydrogen ions and dissipates the hydrogen ion gradient. This leads to a fall in ATP production. (This is also covered on page 90.)

✔ *Quick check 5*

Signals from nearby cells may be involved. These include:
- cytokines
- hormones
- growth factors.

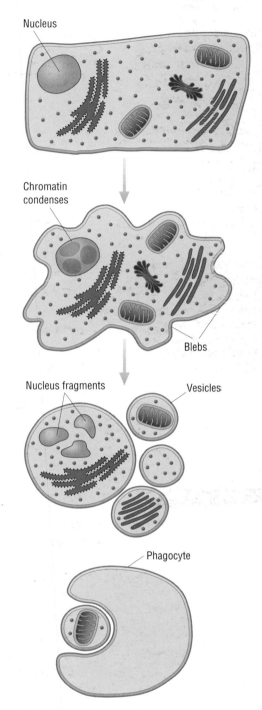

Nucleus

Chromatin condenses

Blebs

Nucleus fragments

Vesicles

Phagocyte

Series of events in apoptosis

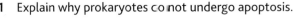

QUICK CHECK QUESTIONS

1 Explain why prokaryotes do not undergo apoptosis.
2 Explain why apoptosis is considered to be a homeostatic mechanism.
3 Because apoptosis happens quickly, it is hard to be sure that a dying cell has undergone apoptosis and not necrosis. During necrosis there is no fragmentation of the nuclear DNA. Suggest a way that scientists could tell that cells have undergone apoptosis.
4 Suggest why necrosis, the death of cells due to trauma (injury or toxic chemicals) is not tidy and may lead to an inflammatory reaction.
5 Explain why a reduction in ATP leads to apoptosis.

Cancer

Key words

- cancer
- apoptosis
- tumour
- benign
- malignant
- metastasis
- mutation
- mutagen
- carcinogen
- proto-oncogenes
- tumour suppressor genes
- oncogenes
- TP53 gene
- p53 protein

Key definition

Cancer is the result of uncontrolled mitotic division (cell division).

Normal cell division is controlled by many factors, most of which are genes or their products. If something goes wrong with these control mechanisms then cells fail to stop dividing. They proliferate (multiply) and form a tumour.

Normal cells.

A single cancer cell forms due to exposure to a carcinogen (something that causes cancer, e.g. ultraviolet light, radiation, tobacco smoke) and divides by mitosis.

Cancer cells divide in an uncontrolled way.

Stage 1   **Stage 2**   **Stage 3**

Hint

Cancerous cells do *not* divide any faster than normal cells. They fail to stop dividing when they should (after about 50 cell cycles), but each cell cycle still takes the same time as in normal cells.

As the cancer cells divide, the tumour gets bigger and becomes distinguishable from normal cells.

The tumour develops its own blood and lymphatic vessels which can carry tumour cells to other parts of the body, via lymph nodes and the blood.

The tumour cells squeeze through blood vessel walls in other parts of the body, causing secondary tumours. This process is called metastasis.

 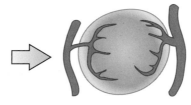 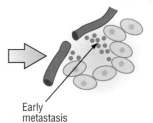

Early metastasis

Stage 4   **Stage 5**   **Stage 6**

How cancer develops

✔ *Quick check 1*

✔ *Quick check 2*

Hint

Oncology is the branch of medicine that involves the study of cancer. If you remember that you will remember that oncogenes cause cancer and proto-oncogenes regulate normal cell division. An oncogene is a tumour-inducing agent.

Proto-oncogenes and oncogenes

Proto-oncogenes help regulate cell division.

- They need to be switched on by hormones or growth factors.
- When switched on they cause cells to move into the S phase of the cycle; once in S phase, the cell is committed to dividing.

If they are mutated they become **oncogenes**.

- Oncogenes are defective proto-oncogenes: they are permanently switched on and do not need signals, such as hormones, from other cells.
- If proto-oncogenes over-express and make too much of their protein they can cause cancer, so they are called oncogenes when doing this.

✔ *Quick check 3*

Some causes of cancer	Further detail
Failure of cells to undergo **apoptosis**	• Cells continue to divide and form a **tumour** • If the tumour stays in one part of the body it is **benign** • If cells break away from it, enter the blood or lymph system and colonise other tissues, then the tumour is **malignant** • The spreading of its cells is called **metastasis**
Mutation of genes that control cell division The mutation may occur in **proto-oncogenes** or **tumour suppressor genes**	**Mutagens** cause mutations. A carcinogen is a mutagen that causes a mutation of genes controlling cell division causing cancer Examples include: • ultraviolet (UV) radiation in sunlight • ionising radiation, such as X-rays • chemicals in tobacco smoke, such as benzopyrene • alcohol
Viruses	Some viruses can also interfere with genes that control cell division, and so can cause cancer. An example is the human papilloma virus that causes cervical cancer

Tumour suppressor genes

These inhibit the cell cycle.
- They code for proteins that stop mitosis. They may do this by repressing other genes that promote the cell cycle. Stopping mitosis causes the cells to undergo apoptosis.
- They couple the cell cycle to DNA damage. They stop cell division so that damaged DNA can be repaired. If the DNA can't be repaired then the cell undergoes apoptosis.

Usually, both copies of a tumour suppressor gene need to be mutated for cancer to occur. (Remember, cells are diploid.)

TP53 is a tumour suppressor gene that makes a protein, **p53**.
- If TP53 is mutated, it makes an altered p53 protein.
- This altered protein fails to stop mitosis, so the cell cycle continues; apoptosis does not happen and a tumour forms.

Benzopyrene and UV radiation will mutate TP53. TP53 will lead to cancer even if only one of its alleles is mutated.

✔*Quick check 4*

QUICK CHECK QUESTIONS

1 Why are cancer cells described as 'immortal'?
2 Suggest why cells are committed to dividing, once they have entered the S phase of the cell cycle.
3 Suggest why oncogenes are also known as 'anti-apoptosis genes'.
4 Suggest why TP53 is an important tumour suppressor gene.

UNIT 2

Stem cells and cell differentiation

Key words

- differentiation
- stem cell
- bone marrow
- pluripotent
- multipotent
- erythrocytes
- leucocytes

Key definitions

Differentiation is the process that stem cells undergo to become specialised or differentiated into a specific cell type. Differentiation involves switching some genes off as the cell matures, whilst ensuring that some are able to be switched on to make the proteins that the differentiated cell needs to perform its specific functions.

Stem cells are undifferentiated cells that are capable of dividing and becoming any type of cell in the organism from which they were obtained.

Sources of stem cells

Embryonic stem cells	Umbilical cord blood stem cells	Adult stem cells
Early-stage embryos, of about 4–5 days after fertilisation, are called blastocysts	Can be obtained from the umbilical cord immediately after birth and frozen	**Bone marrow** contains blood stem cells and other stem cells
They consist of 50–150 cells that are **pluripotent** – capable of differentiating into any one of the 220 or more cell types in the human body. They are not totipotent	May be used for bone marrow transplants	In November 2008 a trachea transplant was carried out in Barcelona. Stem cells derived from the patient's bone marrow were used to make cartilage cells to coat a donated trachea. Epithelial cells from the recipient were used to coat the inside of the trachea, and cartilage cells, developed from her bone marrow stem cells, coated the outside
However, this source of stem cells raises ethical issues as some people believe that every embryo is a potential human	Use does not raise the same ethical dilemma as embryonic stem cells	
	Cells can be stored in cord blood bank to use for treatment of child in later life	This 'tricked' her immune system to treating the trachea as hers and so she did not need any immunosuppressant drugs

✓ *Quick check 1*

Adult stem cells are described as **multipotent** as there are a limited number of cell types that can be developed from them.

However, pluripotent adult stem cells have been derived from fibroblast (flat undifferentiated cells found in connective tissue) cultures.

Scientists hope to use stem cells in regenerative medicine to treat diseases such as:

- Type 1 diabetes
- Parkinson's disease
- blindness due to optic nerve or brain damage
- spinal cord injuries.

✓ *Quick check 1*

Organs or tissues grown from stem cells could be used to test the effects of new drugs, possibly eventually eliminating the need to test drugs on animals.

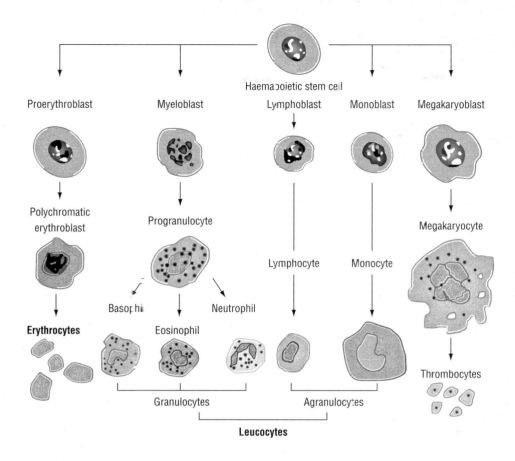

Blood cell production in the bone marrow

Important events in stem cell research:

- In 2003, adult stem cells were isolated from children's milk teeth.
- In 2007 at Newcastle University, scientists grew an artificial liver from stem cells.
- In January 2008, human embryonic stem cells were obtained and cultured without destroying the embryo from which they were obtained. In the same way that pre-implantation genetic screening is carried out, a cell can be taken from an eight-cell-stage embryo, and the embryo can still survive and develop normally.
- In March 2008, a successful operation to regenerate cartilage in a knee, using adult stem cells, was carried out.
- In October 2008, embryonic-like stem cells were obtained from a single human hair.

✔*Quick check 2*

✔*Quick check 3*

QUICK CHECK QUESTIONS

1 Explain the following terms: pluripotent; multipotent; stem cell; differentiation.

2 Discuss possible uses of an artificial liver grown from stem cells.

3 The human embryonic stem cells obtained from 8-day-old embryos were totipotent. This means they each had the potential to become a new embryo. However they were not used for this, but to grow stem cells. Discuss the pros and cons of obtaining human embryonic stem cells from embryos in this way.

Cancer – risk factors and methods of detection

Key words

- epidemiological
- prevalence
- incidence
- menopause
- breast cancer
- BRCA1 gene
- ionising radiation
- mutations
- carcinogen
- ageing
- viruses
- heredity
- diet and bowel cancer
- smoking and lung cancer
- X-rays
- mammography
- CT scans
- thermography
- ultrasound
- MRI
- PET scans

Hint

All cancers are genetic as they involve damage to DNA. However, not all cancers are hereditary.

✔ *Quick check 1*

Key definitions

Epidemiology is the study of factors that affect the health and illness of populations. It informs policy makers on what interventions to make regarding public health and preventive medicine. It is important in evidence-based medicine to identify risk factors for diseases.

Prevalence refers to the number of people with a disease in a given population over a given time. **Incidence** would be the number of new cases. The prevalence of breast cancer is higher in post-**menopausal** women but their survival rate is higher than for younger women because it is easier to diagnose in older women, so is more likely to be diagnosed early. About 1% of **breast cancer** cases are in men and between 5% and 10% of breast cancer cases are hereditary.

The table shows some of the factors that can increase the risk of developing cancers.

Risk factor	Type of cancer	Other information
Ionising radiation	Skin Many types	Ionising radiation consists of high energy, short wavelength rays, such as X-rays, gamma rays and ultraviolet light. DNA can be repaired but **mutations** accumulate
Carcinogens, e.g. Asbestos Aniline dyes Benzopyrene Alcohol	 Lung Bladder Lung Mouth, throat, oesophagus, liver, breast	Prolonged exposure increases the risk
Ageing	Most	Testicular cancer and leukaemia are more prevalent in younger people. Most cancers are more prevalent in older people as there has been more time for mutations to accumulate
Viruses	Leukaemia Cervical	Viruses can trigger changes in the host DNA
Heredity	5–10% of breast cancers	Genes, such as BRCAI, which are linked to breast cancer can be screened for Some types of colon cancer are also hereditary
Diet	35% of all cancers	High fat intake increases risk of breast cancer Low fibre diets increase risk of colon cancer

There is sufficient epidemiological evidence to link **lung cancer** with smoking tobacco (and with passive smoking). Fifty years after the first epidemiological studies, the governments of many countries banned smoking in public places. There is also ample evidence that low fibre diets increase the risk of **bowel cancer**.

The majority of cancers are caused by lifestyle factors, such as smoking, diet (lack of fruit and vegetables for fibre and antioxidants, and high fat content), drinking too much alcohol, and having no children or having children later in life (this is linked to breast cancer).

Compelling evidence

- Bowel cancer risk is 33% higher in people who eat more than two portions of red or processed meat a day.
- There is a strong correlation between development of lung cancer and both the length of time a person has smoked and the number of cigarettes a day smoked. The first study in 1950 showed a 26-fold increased risk of lung cancer in smokers of 15–24 cigarettes a day compared to non-smokers.

✓ *Quick check 2*

Methods of detecting cancer

Method	How it works
X-rays	Tumours are dense and show up as white as fewer X-rays pass through them
Mammography	Low dose X-rays pass through breast tissue and a tumour shows up as white
CT (computed tomography) scans	X-ray pictures are taken from all angles and a three-dimensional image is built up of the inside of the body
Thermography	Tumours are warmer and show up as red/orange/yellow
Ultrasound	Sound waves are used to build an image, particularly useful for tumours in soft tissue, such as the liver
MRI (magnetic resonance imaging)	Measures the magnetic alignment of hydrogen atoms in the body. Tumours have a lot of water and many hydrogen atoms. This scan gives a very clear (high resolution) picture
PET (positron emission tomography) scans	Gamma rays, emitted from an injected radioactive isotope, are detected and an image is built up. The isotope can indicate which cells are dividing/more metabolically active

✓ *Quick check 3*

✓ *Quick check 4*

QUICK CHECK QUESTIONS

1 Explain why all cancers are described as genetic but not all are hereditary.

2 What advice would you give someone to help them reduce their risk of developing cancer?

3 Suggest why MRI scans are considered a safer way to screen women for breast cancer than using X-rays or CT scans.

4 Suggest why tumours contain a lot of water.

Key words

- chemotherapy
- immunotherapy
- clinical trials
- randomised double blind trial
- placebos
- NICE
- cost-effective
- lumpectomy
- mastectomy
- lymph
- chemotherapy
- Tamoxifen
- Herceptin
- immunotherapy
- complementary therapies
- social, ethical and economic considerations

Key definitions

Chemotherapy is the use of chemicals to kill cells – either microorganisms (using antibiotics) or cancer cells.

Immunotherapy is the modulation of the immune system for prophylactic or therapeutic treatment of disease.

New drugs and treatments have to be tested before they can be used to treat patients. A drug being developed is:

- tested, in a lab, on cell cultures or on animals
- then used in **clinical trials**.

These are carried out in four phases, to find out

1 if they work
2 what dosage is needed/safe
3 if there are any side effects
4 what the long-term benefits are (phase 4 is carried out after the drug has been licensed and is being prescribed).

When the drug has been in use for some time, comparative studies compare its efficacy with that of other drugs.

Phase 3 trials are **randomised** and often are **double-blind**. A large group of people agree to take part. Half of them are given the new drug and the others are given an existing treatment or a **placebo**. Who gets what is randomly decided by computer, and the patients and doctors do not know who is in which group until the end of the study. This avoids bias.

NICE (National Institute for Clinical Excellence) decides whether a drug is **cost effective** and whether it should be prescribed by the NHS. Some cancer treatments are expensive but the total cost of treating all cancer patients is very low compared to the cost to the health service of treating obesity or diabetes.

The table on the next page outlines some treatments for cancer.

✔ *Quick check 1*

Hint

The design of drug trials is something you would be expected to comment on as an aspect of 'How Science Works'. You may be asked to comment on the validity and reliability of evidence from trials.

Age	Risk
Up to age 25	1 in 15 000
Up to age 30	1 in 1900
Up to age 40	1 in 200
Up to age 50	1 in 50
Up to age 60	1 in 23
Up to age 70	1 in 15
Up to age 80	1 in 11
Up to age 85	1 in 10
Lifetime risk (all ages)	1 in 9

Breast cancer risk and age

Treatment	Brief description and evaluation
Surgery (breast cancer) • **lumpectomy** • **mastectomy**	 Removal of small area of breast tissue – the lump. Higher rate of recurrence Removal of one or both breasts. Sometimes prophylactically in high risk groups. Used for breast cancer in men. Radical mastectomy involves removal of **lymph** nodes as well. May be followed by breast reconstruction surgery or use of a prosthesis
Chemotherapy	Targets dividing cells. Often used with other treatments. Used to: • shrink tumour before surgery • shrink tumour so immune system can deal with it • give palliative care (prolong life expectancy whilst not curing) Some agents kill cells; some arrest the cell cycle by preventing DNA synthesis, inhibiting spindle formation or preventing chromatid separation in anaphase
Tamoxifen (breast cancer)	This is the generic name for drugs such as Nolvadex, Istabil and Valodex. It is an oestrogen-receptor antagonist and starves tumours of oestrogen, which some need to grow. Used to treat early breast cancer in pre-menopausal women. May be used with cytotoxic (kills/poisons cells) chemotherapy. Side effects include increased risk of uterine cancer and fatty liver. Its action may be blocked by some antidepressants
Herceptin (breast cancer)	This is a monoclonal antibody that acts on the HER2/neu receptors. Used to treat tumours where cells over-express HER2 growth factor and would otherwise keep dividing. It arrests them in G_1 of the cell cycle
Immunotherapy	1 Vaccine (prophylactic) against human papilloma virus available to women aged 12–26 years to prevent cervical cancer 2 Interferon – stimulates T killer cells to kill tumour cells 3 Bacillus Calmette-Guérin (BCG) – used to treat early stage bladder cancer 4 Dendritic cells (antigen presenting cells), taken from patient, are exposed to patient's tumour antigens; they then, on replacement in body, stimulate T killer cells to destroy (cancer) cells with those antigens
Complementary therapies	Some are holistic and help patient to feel better during treatment Gerson treatment involves a strict diet and exercise regime and has been used to treat bone cancer

✔ *Quick check 2*

✔ *Quick check 3*

✔ *Quick check 4*

There are many **social, ethical and economic considerations** involved in developing, testing and administering treatments for cancer patients.

QUICK CHECK QUESTIONS

1 Suggest why placebos are not used in trials for cancer treatments.

2 Discuss the social, ethical and economic issues of a young woman, with a family history of breast cancer, deciding to undergo prophylactic mastectomy.

3 Many chemotherapeutic agents come from plants. Vincristine and vinblastin come from the Madagascar periwinkle. Taxol comes from yew. Suggest why rain and cloud forests are important for cancer therapy.

4 Explain, with reference to examples, what immunotherapy is.

Meiosis and its significance

Key words

- chromosome
- haploid (n)
- gametes
- alleles
- crossing over
- chiasma (plural: chiasmata)
- genetic variation
- independent assortment
- chromatid
- interphase
- prophase 1
- metaphase 1
- anaphase 1
- telophase 1
- prophase 2
- metaphase 2
- anaphase 2
- telophase 2

Key definition

Meiosis means 'a reduction'. This is the type of nuclear division, in eukaryotes, where one diploid cell produces four daughter, haploid cells (each having half the number of chromosomes). The daughter cells are genetically different from each other and from the parent cell.

The significance of meiosis

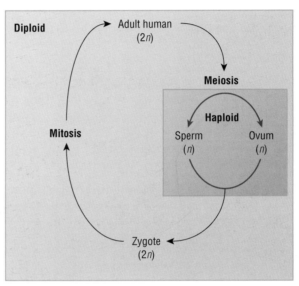

The importance of meiosis

Examiner tip

Make sure you always spell *meiosis* correctly. *Meiosis* and *mitosis* are similar words with very different meanings and there must be no doubt in the examiner's mind as to which you are talking about.

Hint

Individuals don't adapt to changes in the population and *individuals* don't evolve. Some individuals are genetically different and can survive the changed environment. They survive and breed and pass on those favourable alleles, producing an adapted **population**. This is natural selection.

- It halves the number of **chromosomes**. If, as in humans, the resulting **haploid (n)** cells are **gametes** (sex cells, see page 100), then two can join to restore the full (diploid [2n]) number of chromosomes in the resulting zygote. The random fusion of the two gamete nuclei, from (usually) unrelated individuals, at fertilisation, gives a new combination of **alleles** in the individual that develops from the zygote. ✓*Quick check 1*

- In prophase I, homologous chromosomes, each consisting of a pair of chromatids, pair up and undergo **crossing over**. This means that non-sister chromatids swap alleles. The actual places where the chromatids cross over and temporarily attach to each other are called **chiasmata** (sing.: chiasma). Crossing over produces **genetic variation** in the resulting gametes by producing new combinations of alleles. ✓*Quick check 2*

- Homologous pairs of chromosomes attach to the equator of the spindle in metaphase 1 (metaphase of meiosis 1). The orientation of the maternal and paternal chromosomes in each homologous pair (bivalent) is random and also determines how the **chromosomes independently assort** (separate) at anaphase 1. There are 2^n different possibilities. In humans n = 23, so you can see that this results in a lot of genetic variation in the gametes.

- In metaphase 2 (metaphase of meiosis 2) the pairs of **chromatids** line up randomly on the spindle equator and the way they attach determines how they independently assort (separate) at anaphase 2. This further increases genetic variation in the gametes.

Why is genetic variation important?

Genetic variation aids survival of populations and drives evolution. It allows populations of a particular species to adapt to changes in the environment. This means that whatever the change, some individuals will survive. If a population within a species accumulates enough genetic differences so that they can no longer interbreed with the original species, then it becomes a new species.

Division I

Prophase I

Metaphase I

Anaphase I

Telophase I

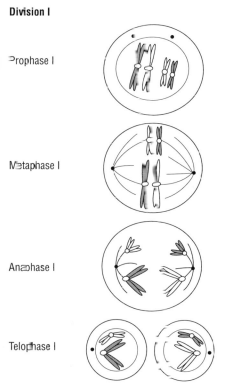

Division II

Prophase II

Metaphase II

Anaphase II

Telophase II

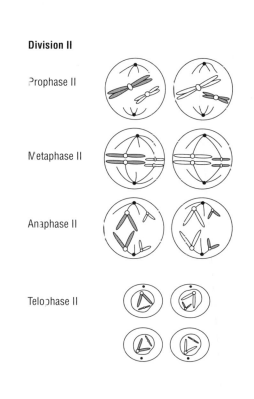

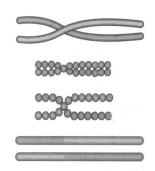

Crossing over occurs in prophase 1. You can see that the chromatids break and then exchange pieces, so that new combinations of genetic information are formed.

The events of meiosis. Notice that there are two divisions, each of which has four stages.

✔*Quick check 4*

QUICK CHECK QUESTIONS

1 Name two organs in humans where meiosis takes place.

2 What are homologous chromosomes?

3 Complete the table opposite to compare mitosis and meiosis.

4 Describe the events taking place in the stages of meiosis, as shown in the diagram.

	Meiosis	Mitosis
Number of divisions		
Products		Two genetically identical daughter cells
Chromosome number		Maintained
Do bivalents form?		
Does crossing over occur?		

Preconceptual and antenatal care during pregnancy

Key words

- antenatal care
- postconceptual care
- preconceptual care
- immune status

Key definition

Ante means before and **natal** means birth, so **antenatal care** (also called **postconceptual care**) is the healthcare given to a woman during her pregnancy, before she gives birth.

Care before pregnancy

Before a woman becomes pregnant she should take care of her health. This is **preconceptual care**. Poor health behaviours such as smoking, binge drinking and not eating a balanced diet before conception can jeopardise the fertility of the mother (and father) and the health of the baby she is about to conceive.

✓Quick check 1

It is also important that all women are **immune** to rubella (German measles) as the virus that causes this infection can cross the placenta and cause birth defects. All girls should be immunised at an early age.

✓Quick check 2

Care during pregnancy

As soon as a woman becomes pregnant she should contact her midwife or general practitioner (GP) so that they can organise her antenatal care. At her 12-week appointment they check:

- her general health – Does she suffer from a chronic disease such as diabetes or epilepsy?
- her history – Has she had any babies with abnormalities or any pregnancies with complications?
- her family history – Are there any genetic diseases in the family?
- previous pregnancies – Has she had any miscarriages?

They organise her:

- screening tests – such as for alpha fetal protein that indicates the risk of the baby having Down's syndrome and which has to be done within a specific time frame
- ultrasound scans to check that the baby is developing and growing normally. These scans can also inform the mother of the sex of her unborn child, if she wants to know.

At subsequent antenatal clinics the following are checked:

- mother's weight and blood pressure

✓Quick check 3

- her urine composition (some women develop diabetes during pregnancy)
- the baby's position and growth.

At antenatal classes the mother meets other mums, and learns about:

- the birth process
- relaxation exercises
- how to care for the baby
- how to care for herself (diet and exercise) during and after pregnancy.

Partners may attend some of these classes. Antenatal classes may be run by a midwife, a GP, a health centre or the National Childbirth Trust (NCT).

The mother may see:
- an obstetrician (a doctor who specialises in pregnancy, labour and the immediate postnatal period)
- a physiotherapist (who gives advice on exercises during and after pregnancy)
- a dietician (who gives advice on diet)
- a health visitor (who meets the whole family)
- a sonographer (a member of the radiography team who carries out the ultrasound scans).

Antenatal visits become more frequent as the pregnancy advances, and the mother's mental and emotional health are also assessed.

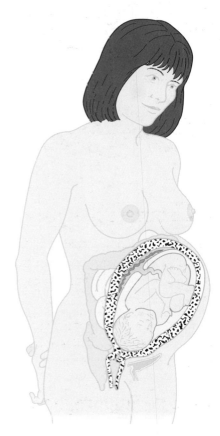

Position of fetus during late pregnancy

Routine tests carried out during pregnancy

Test	Reason for test
ABO blood group	In case the mother needs a blood transfusion after the birth
Rhesus blood group	A Rhesus-negative mother carrying a Rhesus-positive baby needs to be desensitised so that she does not make antibodies against the blood of other babies she may have
Hepatitis virus	Hepatitis B virus can be passed on during childbirth. It can cause chronic liver disease and other problems such as liver cancer
HIV	This virus can pass across the placenta or in breast milk. Medical treatment can reduce the risk of this infecting the baby
Syphilis	This bacterium can affect the baby
Immunity to rubella	If the mother is not immune she will be carefully monitored as she cannot be vaccinated during pregnancy as the vaccine contains live viruses

QUICK CHECK QUESTIONS

1   Suggest why a poor diet and/or heavy alcohol consumption and smoking can affect fertility or harm a baby not yet conceived.

2   Suggest why it is important for girls to be immunised against rubella, long before they are likely to become pregnant.

3   If a woman has pregnancy diabetes there is a risk that her baby can grow too large in the uterus. Why is this?

UNIT 2

Diet during pregnancy, and monitoring fetal growth and abnormalities

Key words

- growth
- protein
- calcium
- vitamins A and C
- folic acid
- iron
- DRV (dietary reference values)
- energy
- nutrients
- alcohol
- carbon monoxide
- nicotine
- ultrasound
- crown-rump length of back
- biparietal diameter

✓ *Quick check 1*

Key definition

Growth involves cell division and an increase in mass.

Dietary advice

Pregnant women should eat plenty of fresh fruit and vegetables as part of a healthy balanced diet, with slightly increased energy content. They should eat enough:

- **protein** for the baby, uterus and placenta to grow
- **calcium** for bone mineralisation
- **vitamin C** for collagen formation
- **vitamin A** for making epithelial tissues and a pigment, rhodopsin, in rod cells of the eyes, but no more vitamin A than usual as it is a teratogen and can cause birth defects if taken in large amounts
- **folic acid** (vitamin B9) to prevent neural tube (brain and spinal cord) defects
- **iron** for haemoglobin formation
- fish oils (long chain omega 3 fatty acids) for brain development.

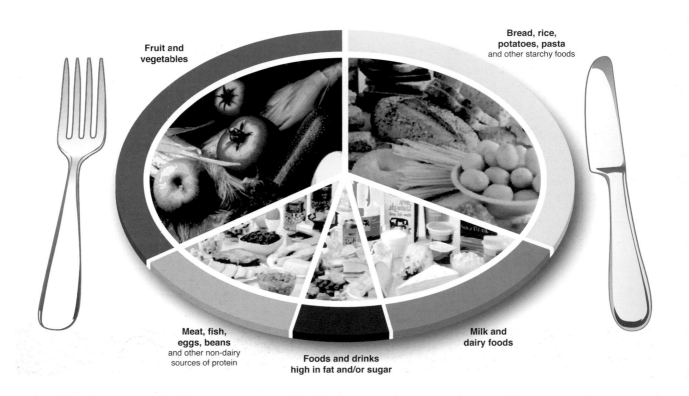

This figure shows how much of what you eat should come from each food group

Dietary reference values (DRV) are published to give guidelines as to how much **energy** and **nutrients** certain groups of people need. The estimated average requirement (EAR) is enough to meet the needs of 50% of the population. RNIs (reference nutrient intakes) indicate the amounts needed to meet the needs of 95% of the population.

What to avoid

- **Alcohol** – it is a teratogenic drug, which means it interferes with gene action and expression. In excess it causes birth defects, heart and facial deformities, and impaired brain development (fetal alcohol syndrome).

✓ *Quick check 1*

- Smoking – **carbon monoxide** combines irreversibly with haemoglobin and reduces the amount of oxygen reaching the fetus so its respiration, and therefore its growth, are reduced. **Nicotine** may also reduce birth weight, as it can constrict blood vessels and reduce nutrient delivery to the fetus and it may also decrease the mother's appetite.

✓ *Quick check 2*

- Recreational drugs – these can alter the fetal metabolism and cause addiction in the baby.
- Many medicinal drugs – these may harm the baby.
- Children with virus infections – cytomegalovirus (CMV) can cross the placenta.
- Liver, as it contains a lot of vitamin A.

✓ *Quick check 3*

- Unpasteurised cheeses as they may contain *Listeria* – a bacterium that can increase the risk of miscarriage.
- Undercooked meat or eggs as they may contain *Salmonella* bacteria.

✓ *Quick check 3*

- Swordfish, fresh tuna, shark or marlin as they may contain mercury.
- Putting on too much weight.

Measuring fetal growth

A sonographer uses **ultrasound** scans. Sound waves are reflected back from the body to produce an image on a screen. The mother's bladder must be full and a gel is put on her abdomen so there is no air between the instrument and the body.

The **crown-rump length** of the fetus is measured from the image and, knowing the stage of the pregnancy (calculated from the date of the mother's last menstrual period), the sonographer and obstetrician can tell if the fetus is growing normally.

The diameter of the head at its widest point, the **biparietal diameter**, is also measured from 12 weeks onwards to see if the brain is developing normally.

From 18 weeks onwards most of the organs of the fetus can be seen and the scan shows if these are developing normally.

QUICK CHECK QUESTIONS

1 Explain, with reference to examples, what 'teratogen' means.
2 Explain why pregnant women should not smoke tobacco.

3 Suggest why pregnant women should avoid eating certain foods, such as (a) liver, (b) soft unpasteurised cheese, and (c) mayonnaise made from raw eggs.

UNIT 2

Chromosomal mutations and the use of karyotyping

Key words

- mutation
- chromosome
- trisomy
- non-disjunction
- Klinefelter's syndrome
- Turner's syndrome
- karyotype
- amniocentesis
- chorionic villus sampling
- ultrasonography

Key definition

A **mutation** is a change to genetic material. Some mutations involve a change to a piece of DNA and some involve changes to **chromosome** structure.

Chromosome mutations may cause a change to the *structure* of chromosomes, such as loss or duplication of part of a chromosome or the transfer of part of one chromosome to another chromosome.

Chromosome mutations may also result in changes to the *number* of chromosomes.

Structural abnormalities	Numerical abnormalities
- duplication of a chromosome or of part of a chromosome - loss of part of a chromosome - inversion – a piece of a chromosome is removed, turned through 180° and then reinstated at the same position on the chromosome - translocation – a piece of one chromosome attaches to another chromosome	- Extra chromosome **Trisomy** means having an extra copy of one chromosome. This happens if one pair of chromosomes fails to separate properly at anaphase 1 of meiosis or chromatids of a pair fail to separate at anaphase 2 of meiosis. This failure to separate is called **non-disjunction**. Of the resulting gametes, some will have an extra chromosome and some will lack a chromosome. If one with 24 chromosomes joins with a normal gamete, the zygote will have an extra chromosome, e.g. Trisomy 21 (Down's syndrome) Trisomy 18 (Edwards' syndrome) XXY (**Klinefelter's syndrome**) XXX XYY - Lacking a chromosome About 1 in 6000 females is born lacking one of the X chromosomes (monosomy). A normal gamete has joined with one that lacks a sex chromosome. The genotype for the sex chromosomes is written as XO. **Turner's syndrome** females are short, have a webbed neck, shield-shaped chest, underdeveloped breasts, may not menstruate, and have cardiovascular and renal problems

✔ *Quick check 1*

✔ *Quick check 2*

Karyotyping

A **karyotype** is a photomicrograph of chromosomes from metaphase of mitosis, with the chromosomes arranged in a standard sequence, so doctors can see if there is an abnormality of the number of chromosomes in the fetus.

To prepare a karyotype, fetal cells have to be obtained, by **amniocentesis** or **chorionic villus sampling**.

Test	Amniocentesis	Chorionic villus sampling
Procedure to obtain cells	Hollow needle inserted into amnion in pregnant uterus; amniotic fluid with sloughed off fetal skin cells withdrawn Ultrasound scan (**ultrasonography**) carried out so needle does not damage the fetus or placenta	Tube inserted into vagina, through cervix and into uterus Ultrasound scan used to show where placenta is Cells from fetal part of placenta obtained
Advantages	Enables parents to prepare for a child with an abnormality or to terminate the pregnancy	Can be done from 10 weeks, so if parents decide to terminate the pregnancy it may be less traumatic
Disadvantages	Cannot be done until the 15th week as before that the amniotic sac is not large enough 1% risk of causing a miscarriage	2% risk of causing a miscarriage

Making the karyotype

1 Cells are cultured, under aseptic conditions, for about 3 weeks, so they are dividing by mitosis.
2 Cells are arrested at the beginning of metaphase by adding colchicine, which inhibits spindle formation.
3 The cells are placed in a solution of very high water potential, so that they swell and burst and the chromosomes spread out from each other.
4 A stain is added to make the chromosomes easily seen.
5 These cells are observed under the microscope and a photograph is taken.

✓*Quick check 3*

✓*Quick check 4*

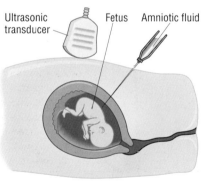

Amniocentesis

Ultrasonic transducer — Fetus — Amniotic fluid

QUICK CHECK QUESTIONS

1 Sometimes, a zygote forms with three sets (69) of chromosomes. This is not viable and it spontaneously aborts. Suggest how a zygote with 69 chromosomes can occur.

2 Some patients with Turner's syndrome are found to have some cells with X0 and some with XX or XY chromosomes. Such patients are called mosaics as they have two cell lines in their bodies. What does this suggest about when the non-disjunction of chromosomes occurred in these patients?

3 Karyotyping could show a chromosome abnormality and prospective parents may decide to terminate the pregnancy. What are the risks to be considered before opting for such a test?

4 Explain why human cells placed in very dilute salt solution will swell and burst.

Growth patterns during the human life cycle

Key words

- life cycle
- absolute growth rate
- relative growth rate
- mass
- length
- height
- head circumference

Key definition

Absolute growth rate is the change in mass or height divided by the time period. If a child grows 6 cm in 1 year the absolute growth rate is 6 cm year^{-1} or 0.5 cm month^{-1}.

Key definition

Relative growth rate is a specific growth rate. It shows the increase compared to the initial height. If two girls grow 6 cm in a year, but one is 75 cm at the beginning of the year and the other was 150 cm at the beginning of the year, the first girl has a greater relative growth rate.

$$\text{relative growth rate} = \frac{\text{change in height in one time period (e.g. 1 year)}}{\text{height at beginning of the time period}}$$

$$\text{relative growth rate} = \frac{\text{absolute growth rate}}{\text{height at beginning of the time period}}$$

✔ *Quick check 1*

The human life cycle covers the period from conception, through:

- the fetal growth phase
- infancy
- adolescence

✔ *Quick check 2*

- adulthood.

An individual grows until about age 16–18 years for females and 22 years for males. A graph of absolute growth shows the height or mass over time, and absolute growth rate shows the increase in height or mass in cm per year or kg per year.

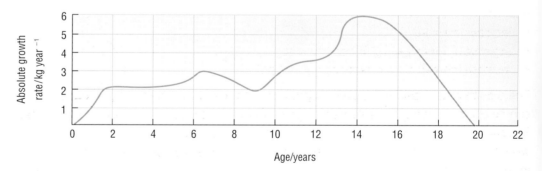

Absolute growth curve for humans

You can see from the first graph that at different periods the growth rate varies. It is very rapid during the first 2 years, then slows, and then shows two growth spurts, at around age 5–7 years and at adolescence, from age 9 to 15 years. The actual timing of the spurts will vary from individual to individual and is different for boys and girls.

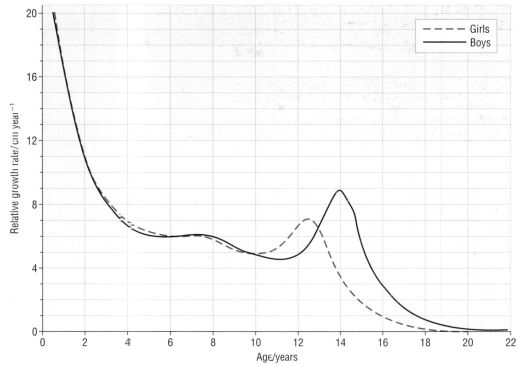

Relative growth rate of boys and girls

The graph shows the relative growth rates: height gained as a percentage of original height each year. You can see that:

- the rate of growth is highest during the first year
- the growth rate slows – the infant is still increasing in mass and height, but more slowly
- there are two growth spurts, one at age 6–8 years and one at 10–15 years
- the adolescent growth spurt is earlier for girls than for boys.

✔ *Quick check 3*

Measurement of growth in infants

We have seen that fetal growth is monitored during pregnancy.

After the birth health visitors continue to monitor each infant's growth. They keep records of the baby's

- increase in **mass**
- increase in **length** – top of head to feet
- increase in **head circumference** (widest point).

The growth rate of children is also checked – height and mass – and compared with centile charts showing the accepted range of 'normal' growth. If a child's height is on the 3rd centile, for every 100 children of that age, three would be the same or shorter and 97 would be taller. These charts were based on a group of mainly bottle-fed white infants in the USA. They have probably caused health visitors to encourage mothers to overfeed their babies. This could have contributed to the current rise in obesity among children.

✔ *Quick check 4*

QUICK CHECK QUESTIONS

1 Explain the difference between absolute and relative growth rates.

2 List the four phases of human growth.

3 Explain why, in a class of 13-year-old boys and girls, many of the girls are taller than the boys.

4 What criteria do you think should be used to make new infant growth charts?

Differential growth in humans and maintaining healthy growth

Key words

- multicellular
- tissues
- organs
- systems
- nervous tissue
- reproductive tissue
- lymphatic tissues
- carbohydrates
- essential fatty acids
- lipids
- essential amino acids
- proteins
- calcium
- iron
- phosphorus
- vitamins A, C and D

Key definition

Growth is an increase in size of an organism by an increase in its body substance. It involves cell division, and as humans grow all their organs increase in size. Growth in humans is coordinated by various hormones

Hint

Cells are organised into:

- **tissues** (a collection of similar cells that are adapted to perform a specific function, such as epithelial tissue)
- which are organised into **organs** (collections of tissues that perform a specific function or functions, such as the heart)
- which are organised into **systems**, such as the circulatory system.

Different parts of the body grow at different rates

Humans are **multicellular** organisms. During embryonic development, cells have become differentiated and specialised, as some genes are switched off and others switched on.

Brain and nervous tissue	Reproductive organs	Lymphatic (immune) system
Brain in the newborn is underdeveloped so head is not too large for birth canal	Start to develop at puberty	Underdeveloped at birth
Brain and nervous system develop rapidly in first 5–7 years	By adulthood humans can reproduce and pass on their genes	Antibodies pass from mother across placenta and in breast milk, and provide temporary immunity
✓*Quick check 1*	✓*Quick check 2*	Immune system develops during childhood as children are exposed to pathogens
Brain full size by age 10 years, but neurones continue to develop dendrites to communicate with other neurones and improve mental performance		✓*Quick check 3*

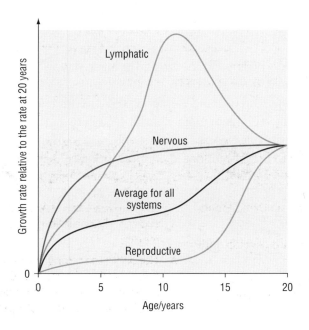

Growth rates of reproductive, lymphatic and nervous systems in humans and overall (average) growth rate

Maintaining healthy growth

Babies and infants need a balanced diet for them to grow and develop properly, and for their bodies to work properly. The table below shows some of the components needed. Children also need fibre and water, as well as vitamins B and E.

Nutrient and Source	Use in body
Carbohydrates Bread, potatoes, rice	Respired to provide energy, e.g. for cell division, active transport, synthesis of molecules Best eaten as starch and not sugar
Essential fatty acids Oily fish	Linoleic and linolenic acid needed to make cell membranes, and for healthy functioning of kidneys, immune system and circulatory system. Fish oils needed for development of brain and nervous tissue. These are essential in diet as humans cannot make omega 3 fatty acids (apart from pregnant and lactating women, who can make them) ✓*Quick check 4*
Lipids Butter, margarine, egg yolk	For insulating nerve cells, making cell membranes, as a source of energy when respired, as an energy store, to cushion organs, as a source of fat-soluble vitamins
Essential amino acids Milk, cheese, egg white, meat, soya	Humans need to take in eight essential amino acids in the diet. The other non-essential ones are made in liver cells that have the necessary enzymes. Children need ten essential amino acids.
Proteins Milk, cheese, eggs, meat, soya, quorn	As a source of amino acids for growth and repair, to make new organelles, cells and tissues. To make muscle proteins, haemoglobin, cell membrane proteins, enzymes, antibodies, collagen. Can be respired if no other substrates available. BUT if proteins are used for respiration then not enough amino acids will be available for synthesis of proteins such as antibodies or enzymes
Calcium Milk, cheese, yoghurt, vegetables	For hardening of bones and teeth, nerve transmission, blood clotting. Good bone density protects against osteoporosis in later life
Iron Green vegetables, meat	For haemoglobin for oxygen transport
Phosphorus Most foods	For healthy bones and teeth. Also to make DNA, RNA, ADP and ATP. To make phospholipids in cell membranes
Vitamin A Liver, carotene	The body turns carotene (carrots, peppers, green vegetables, melon, squash, apricots) into vitamin A. Needed to make rhodopsin in rod cells of eye, to see in dim light. For maintaining epithelial growth. For healthy bone growth in children. Is an antioxidant and protects from cancer
Vitamin C Citrus fruits, green vegetables, potatoes, tomatoes	For formation of collagen. For healthy skin, gums, blood vessel walls, tendons, ligaments and bones (joints)
Vitamin D Made by skin when exposed to UV light Oily fish, liver, egg yolk, milk	Is a hormone. Regulates uptake of calcium from gut into blood and uptake of calcium from blood into bones and teeth Protects against cancer and heart disease

QUICK CHECK QUESTIONS

1 Explain why human infants are so reliant on their parents for a long period.

2 Explain why the reproductive organs do not develop until puberty.

3 There is evidence that the increase in allergies is due to humans having immature immune systems because they are reared in very clean conditions. Explain why such conditions lead to failure of the immune system to mature.

4 The only time during the human life cycle that essential fatty acids can be made is during pregnancy and lactation. What does this suggest about the genes involved in that metabolic pathway?

UNIT 2

Infectious diseases

Key words

- infectious disease
- endemic
- epidemic
- pandemic
- prokaryotic cell
- TB
- HIV
- AIDS

Key definition

An **infectious disease** is caused by an infecting agent (pathogen), such as a virus or bacterium. They can be spread by direct contact or by indirect means such as infected food or water, or by blood. Some are sexually transmitted in body fluids.

An **endemic** disease is one that is always present in a particular area.

An **epidemic** is a sudden outbreak of a disease that spreads rapidly in an area.

A **pandemic** is an outbreak of a disease that spreads rapidly across continents or across the whole world.

✔ *Quick check 1*

Mycobacterium tuberculosis

This bacterium is a **prokaryote**. It has no nucleus and no membrane-bound organelles. It is rod shaped and about 10 µm long. It causes tuberculosis (TB).

Capsule is a layer of mucilage which may unite bacteria into colonies.

Pili (sing. Pilus) are protein rods for bacteria to attach to host cells or to attach to each other when transferring DNA.

Plasma membrane is a typical phospholipid bilayer.

Plasmids are small pieces of circular DNA which replicate independently of the main genome.

Cell wall has a rigid framework of *murein*, a polysaccharide cross-linked by peptide chains.

Genetic material is composed of a circle of double-stranded DNA *which is not enclosed within a nucleus.*

Ribosomes smaller than those in eukaryotes and not supported on an endoplasmic reticulum.

Food stores are typically lipid droplets and glycogen granules.

Mesosome. This plays a part in cell division and may house enzymes for respiration.

Flagellum is responsible for motility of many bacteria.

SCALE

0.1 µm

✔ *Quick check 2*

The structure of a prokaryotic cell, such as Mycobacterium tuberculosis

The human immunodeficiency virus (HIV)

This is a retrovirus, which means that its genetic material is RNA. When the virus infects a host cell (T helper lymphocyte) its enzyme, reverse transcriptase, makes a DNA copy using the virus RNA as a template. This inserts into the host DNA.

It stays dormant in host immune cells and later causes acquired immune deficiency syndrome (AIDS), making it easier for pathogens to cause opportunistic infections or for cancer tumours to grow, both of which can lead to death.

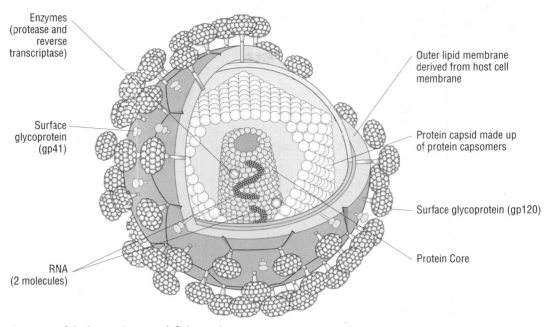

Structure of the human immunodeficiency virus

The symptoms, causes and means of transmission of TB and HIV/AIDS

Disease	Cause	Symptoms	Means of transmission
TB	*Mycobacterium tuberculosis* bacteria In many healthy people the bacteria are present in lungs but encased in macrophages and remain dormant. As people age or become immunocompromised (immune system is weakened), the bacteria become active and cause TB	Fever, weight loss, persistent cough with blood in sputum, fatigue, chest pain, night sweats Usually infects lungs but may infect other organs of the body	Droplet infection Spreads rapidly among people who live in overcrowded conditions and amongst the malnourished ✔ *Quick check 3* Drinking unpasteurised milk from infected cows
HIV/AIDS	Human immunodeficiency virus Infects T helper lymphocytes, macrophages and brain cells, and can remain dormant inside the cells for up to 10 years. When the virus becomes active, the person develops full blown AIDS, often indicated by them succumbing to TB	Mild flu-like symptoms then recovery Stays dormant for up to 10 years May lead to loss of T cells and opportunistic infections, such as TB, toxoplasmosis and cancer	Unprotected sex Across placenta to fetus, or in breast milk from mother to baby Blood to blood as in sharing needles and syringes Use of contaminated blood products

QUICK CHECK QUESTIONS

1. Explain the following terms: (a) endemic, (b) epidemic, (c) pandemic.

2. List the ways in which a prokaryote cell is (a) similar to a eukaryote cell and (b) different from a eukaryote cell.

3. Suggest why malnourished people are more susceptible to infectious diseases.

4. Using your knowledge of how TB and HIV are spread, for each one list the precautions that should be taken to reduce its spread.

UNIT 2

Antibiotics and superbugs

Key words

- antibiotics
- strains
- resistance
- natural selection
- evolution
- MRSA

✓ *Quick check 1*

Examiner tip

Remember, antibiotics are not effective against viruses and viral infections, but they are given to people with viral diseases to treat or prevent secondary infections caused by pathogenic bacteria.

Key definition

An **antibiotic** is a chemical, produced by a microorganism, such as a bacterium or fungus, that can inhibit the growth of or kill other microorganisms, by interfering with their metabolism. Hence, they do not affect viruses as these do not have any metabolism.

In the 1920s Alexander Fleming was culturing bacteria in a lab when he noticed that one plate was contaminated with a fungus, *Penicillium notatum*. He then noticed that no bacteria were growing near the fungus and further investigations led to the discovery of penicillin, a chemical made by that fungus, which inhibits the growth of some bacteria.

Penicillin was made in large quantities during the 1940s and used to treat wounded servicemen during the second world war (up until then, more soldiers died from infections of their wounds than from the actual wounds).

In the 1940s another antibiotic, streptomycin, was available and used for treating tuberculosis (TB). The two antibiotics work in different ways.

The table shows some antibiotics and their modes of action.

Mode of action	Examples
Inhibits cell wall synthesis	Penicillins, cephalosporins, teicoplanin, vancomycin, bacitracin
Inhibits translation stage of protein synthesis	Chloramphenicol, erythromycin, streptomycin, teracyclines
Inhibits transcription stage of protein synthesis	Rifampicin
Inhibits DNA replication	Nalidixic acid, fluoroquinolones
Damages cell membranes	Polymixin B, nystatin, amphotercin B
Inhibits enzyme activity	Sulfonamides

Antibiotics have saved thousands of lives. However, they have been used chaotically (prescribed by doctors when not actually needed and courses not finished by patients) in the past, and such use has caused some populations or **strains** of bacteria to become **resistant** to antibiotics.

In any population of bacteria:
- there will be some with a random mutation such as an altered gene, on the plasmid, that makes the bacterium resistant.
- Normally this altered gene gives the bacterium a disadvantage, but in the presence of the antibiotic these bacteria survive and multiply in the absence of competition from other bacteria and pass on the altered gene.
- This is an example of **natural selection**.
- A new population of bacteria has **evolved**.
- A different antibiotic has to be used to deal with them.

If people take antibiotics when they don't need to:
- resistant bacteria in their gut are selected
- bacteria can swap bits of DNA, so this resistance can be passed on to pathogens.

Some bacteria are resistant to many antibiotics. They are known as 'superbugs'. **MRSA** (methicillin-resistant *Staphylococcus aureus*) is a superbug. (Methicillin is a type of penicillin and is no longer used.) Many people have these bacteria on their skin but if they enter a deep wound it can be fatal. Flucloxacillin, vancomycin and teicoplanin are used to treat MRSA but some strains of *Staph aureus* resistant to flucloxacillin have emerged.

Some strains of TB are resistant to many antibiotics so the treatment involves giving a mixture of antibiotics. The treatment lasts for many months as TB bacteria are slow-growing.

How to reduce the spread of MRSA

Hospitals can adopt methods to reduce the spread of MRSA.

- Isolate all new patients and check to see if they carry MRSA.
- Improve hygiene – hand washing/use of alcohol gels – for all staff, patients and visitors.
- Sterilise any equipment that is used on more than one patient – such as blood pressure monitors.
- Thoroughly clean all wards, equipment, mattresses, beds, curtains, lockers, tables.
- Make staff wear disposable gloves and aprons while examining/treating patients.
- Ban covered lower arms, jewellery, watches buckles, belts, pens and ties.
- Sterilise patients' skin before an operation – MRSA is easily killed by antiseptics.

✔ *Quick check 2*

Examiner tip

Don't say bacteria are 'immune' to antibiotics. They are resistant; it is a totally different mechanism. Bacteria do not have immune systems.

Also, don't say that *people* become resistant to antibiotics. This is not true; it is the *bacteria* that are resistant. However, people may suffer from *side effects* of antibiotics.

✔ *Quick check 3*

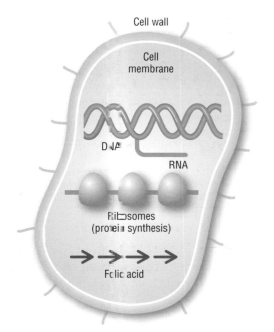

Inhibition of...	
Cell wall synthesis	**Protein synthesis**
Penicillins	Macrolides
Cephalosporins	Chloramphenicol
Carbapenems	Tetracycline
Daptomycin	Aminoglycosides
Glycopeptides	Oxazolidonones
DNA synthesis	**Folic acid synthesis**
Fluoroquinolones	Sulfonamides
	Trimethoprim
RNA synthesis	
Rifampin	

Some of the ways in which antibiotics target and kill bacteria cells

QUICK CHECK QUESTIONS

1 Explain why antibiotics cannot be used to treat viral infections.

2 Explain what is meant by a 'superbug'.

3 Discuss the methods that should be used in hospitals to reduce the spread of MRSA.

UNIT 2

Antimicrobials and other medicines from plants

Key words

- *ex situ* conservation
- rare
- extinct
- seed banks
- drug
- medicinal drugs
- antimicrobial compounds

✓ *Quick check 1*

✓ *Quick check 2*

Key definition

Ex situ conservation is the conservation of **rare** plants or plants that are **extinct** in the wild, because their habitat has been destroyed. These plants are conserved in botanic gardens or in **seed banks**.

Key definition

A **drug** is a substance that alters the metabolism. Hence, it can alter physiology and may affect behaviour. Some drugs are medicinal (medicines) and are legal but need to be used carefully. Caffeine and nicotine are also legal drugs. Some recreational drugs are illegal but may have a medicinal value. Cannabis is a good painkiller (analgesic) for people suffering from multiple sclerosis.

Early humans discovered that certain plants can be used to treat specific ailments. Indeed, some animals eat certain plants, which they don't normally eat, when they are ill. Many modern **medicinal drugs** are derived from plants. The ancient Egyptians used poppy and clay for treating diarrhoea. We still use kaolin (clay) and morphine (an opiate drug derived from poppy seed cases) to treat this ailment.

Some plants produce **antimicrobial compounds**. These chemicals kill or inhibit the growth of bacteria. If these compounds can be linked to certain genetic markers then different plant species can be screened as potential sources of similar compounds.

Plant	Extract	Effect
Garlic	Allicin	Kills some types of bacteria
Artemisia (wormwood)	Artemisinin	Kills malaria parasite
Xylopia aethiopica (Ethiopian pepper)	Extract from the fruit	Kills a range of bacteria and can be used to treat dysentery
Aframomum melegueta (grains of paradise)	Oils obtained from the fruit	Kills bacteria, fungi and schistosome worms that cause Bilharzia

All of the above have been used as remedies in some areas of the world for centuries. Modern scientific research has identified the active ingredients and confirmed their modes of action.

Herbal remedies can also be used to treat conditions such as headaches and depression and have been used to induce abortions. Star fruit reduces hypertension. In Germany all herbalists are also qualified doctors and 75% of all prescribed medicines are herbal in origin.

Chinese herbal remedies for eczema and for ME (myalgic encephalomyelitis; fatigue after a virus infection) appear to be effective.

However, some herbal extracts can be dangerous if taken over a long period of time and they can produce side effects. They may be particularly dangerous if the person is suffering from a condition such as hypertension, so a medical herbalist should be consulted.

Honey and sugar have been used to pack a deep wound to prevent bacterial infection. They kill or inhibit growth of bacteria by causing the cells to lose water by osmosis.

The Chelsea Physic Garden in London was established in 1673, by the Royal Hospital, to train apothecaries to identify useful medicinal plants. There are many species of useful plants there and the garden is open to visitors.

At Kew Gardens is the millennium seed bank. Here, seeds of many rare and threatened plant species are dried and stored so they will remain viable for many decades.

Loss of species

Unfortunately, many plant species in the world are endangered and have even become extinct. Many of the plants that are becoming extinct have not yet been named. It is likely that among the plants being lost are plants containing useful medicines that we do not yet know about. One reason so many plants are becoming extinct is that forests are being cut down to provide timber for buildings, or to clear land to build houses. Forests are also being cleared to create more farmland. Some of the people who are clearing these forests are poor people who are simply trying to make a living for their family.

Some other medicinal drugs obtained from plants:

Drug	Plant	Use
Vincristine Vinblastine	Madagascan periwinkle	For chemotherapy against some types of cancer, e.g. leukaemia
Taxol	Yew	Treats some types of cancer
Aspirin	Willow	Anti-inflammatory and to reduce risk of stroke (it reduces risk of blood clots forming)
Codeine, morphine	Opium poppy	Analgesic
Digoxin	Foxglove	Treats heart failure/atrial arrhythmia
L-dopa	Velvet bean	Treatment of Parkinson's disease
Hypericin	St John's wort (Hypericum)	Treats mild depression but should not be used if taking antivirals or medication for some heart conditions
Phyto-oestrogens	Soya	May alleviate symptoms of menopause

Hint

It is a mistake to assume that because these herbs are 'natural' they will be harmless or better for us than other medicines. After all, many poisons come from natural sources and some plants are very poisonous. Other natural phenomena are also bad for us, such as earthquakes and volcanic eruptions, whereas surgery and vaccination, both 'unnatural' may be very good for us!

✔ *Quick check 3*

Examiner tip

What do we mean by 'bar codes'?

You will meet this idea again in A2 when you look at DNA profiling.

Quite simply, specific regions of DNA are cut up and separated into bands using a technique called gel electrophoresis. This gives a banding pattern. When a pattern is associated, for example, with the production of compounds within medicinal properties, this is a quick way of screening plant species for potential sources of drugs.

QUICK CHECK QUESTIONS

1 Explain what is meant by 'a drug'.
2 What are antimicrobial compounds?

3 Explain why 'natural' substances are not necessarily better for us then 'man-made' substances.

UNIT 2

Vaccination

Key words

- vaccination
- vaccine
- active immunity
- passive immunity
- natural immunity
- acquired immunity
- vaccination programme
- herd immunity

Key definition

Vaccination originally referred to inoculation of patients with cowpox virus (vaccinia), which caused them to be immune to smallpox. The term is now used when antigenic material (the protein antigens of a pathogen or the weakened or dead pathogen) is introduced into the body of a patient to stimulate an immune response with specific antibodies being produced. Afterwards, memory cells (long-lived T and B lymphocytes) remain in the patient, making them immune to the disease caused by that pathogen.

Types of immunity

Hint

The term immunisation is often used as being synonymous with vaccination. Strictly speaking **vaccination** refers to injection of antigenic material to stimulate an immune response. Immunisation refers to injection of antibodies. However, the result of being vaccinated is that you become immune – so you have been immunised.

Type of immunity	Description	Examples
Natural active	Immunity after having been infected with a pathogen causing the body to mount an **immune response**	Memory cells remain in body after an infection of chicken pox Slow acting but long lasting
Acquired active	Immunity after being inoculated with antigen or dead/weakened pathogen, to stimulate an **immune response** Booster often needed as the response is quicker and much greater after second inoculation	Memory cells remain in body after being vaccinated against diseases such as polio, rubella and mumps Slow acting but long lasting
Natural passive	**No immune response** involved Antibodies for antigens to which mother has been exposed are passed to fetus across placenta or to baby in colostrum (first breast milk)	Babies have temporary immunity to diseases mother has been exposed to; this lasts for about a year, whilst the baby's immune system is developing
Acquired passive	**No immune response** involved Antibodies made in a lab are injected	Quick-acting but short-lived, e.g. immunity to tetanus acquired from an anti-tetanus injection after being bitten or having a deep wound infected with soil

✓ *Quick check 1*

Vaccination programmes

'Be wise immunise' was a slogan used to encourage parents to get their children immunised against certain diseases. The table shows the routine childhood immunisation programme in the UK at the present time, but the guidelines can change. For example, 12- and 13-year-old girls are now vaccinated against the human papilloma virus (HPV).

When to immunise	Diseases protected against	Vaccine given
2 months old	Diptheria, tetanus, pertussis (whooping cough), polio and Haemophilus influenzae type b (Hib) Pneumococcal infection	DTaP/IPV/Hib + Pneumococcal conjugate vaccine (PCV)
3 months old	Diptheria, tetanus, pertussis, polio and Haemophilus influenzae type b (Hib) Meningitis C	DTaP/IPV/Hib + MenC
4 months old	Diptheria, tetanus, pertussis, polio and Haemophilus influenzae type b (Hib) Meningitis C Pneumococcal infection	DTaP/IPV/Hib + MenC + PCV
Around 12 months	Haemophilus influenzae type b (Hib) Meningitis C	Hib/MenC
Around 13 months	Measles, mumps and rubella Pneumococcal infection	MMR + PCV
3 years 4 months to 5 years old	Diptheria, tetanus, pertussis and polio Measles, mumps and rubella	DTaP/IPV or dTaP/IPV + MMR
13 to 18 years old	Tetanus, diphtheria and polio	Td/IPV

There are risks associated with most medical treatments, including vaccinations. However, if too few children are vaccinated, then epidemics of infectious diseases can result. Measles is a serious disease and can kill children or leave them brain damaged or deaf. Measles is highly infectious and requires 95% vaccination rates to prevent epidemics.

People have a low understanding of 'risk'. Many will opt not to have their children immunised but still drive them to school each day in the car. About 50 children (and a larger number of adults) die each month in road traffic accidents.

In order to prevent such epidemics, a large proportion of the population has to be vaccinated so that the risk of an infected person passing on the pathogen to a non-immune person is very small. This breaks the chain of infection and stops the disease from spreading.

Herd immunity

About 85–95% of the population needs to be immune for this to happen. Such a state of immunity within a population is called **herd immunity**. As many vaccines are not 100% effective, to gain 95% immunity in the population, just about everyone needs to be immunised.

✓*Quick check 2*

QUICK CHECK QUESTIONS

1   Complete the table to compare active and passive immunity
2   Explain what is meant by 'herd immunity' and why it is important in a population to prevent disease epidemics.

	Active immunity	Passive immunity
Antigen presentation	yes	
Antibodies made by B lymphocytes		
Memory cells made		
Time of action		quick
Duration	long/permanent	

The immune response

Key definition

The **immune response** is a specific response to invasion of the body by a pathogen. Antigens on the pathogen stimulate the production of antibodies and T helper and T killer cells, in the infected body.

The first line of defence

This consists of:
- the skin – tough, dry, salty and inhospitable to pathogens
- mucous membranes of respiratory tract, gut, reproductive tract
- stomach acid – destroys pathogens ingested in food
- ear wax
- tears, containing lysozyme – this enzyme weakens bacterial cell walls.

The non-specific inflammatory response

If **pathogens** get past the first defence lines and enter the body:
- infected cells release histamine
- this increases blood flow to the infected area (hence it feels hot) and makes capillary walls more leaky
- tissue fluid seeps out (so the area swells) and macrophages squeeze out into surrounding tissues
- neutrophils and macrophages (both **phagocytic** white blood cells) ingest and digest the pathogens.

This response is **non-specific**. It is the same for all invading pathogens.

Some macrophages become antigen presenting cells (APCs). They display the pathogen's **antigens** on their cell surface membranes and this stimulates the **specific response**.

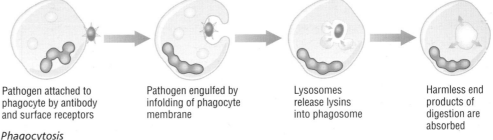

| Pathogen attached to phagocyte by antibody and surface receptors | Pathogen engulfed by infolding of phagocyte membrane | Lysosomes release lysins into phagosome | Harmless end products of digestion are absorbed |

Phagocytosis

The specific immune response

B and **T** cells are both **lymphocytes**. They are made from blood stem cells in the bone marrow. B lymphocytes mature in bone marrow. T lymphocytes mature in the thymus gland.

Clonal selection – the T cell with the corresponding membrane receptor is found by the APC.

- It divides by mitosis to make more of the same T cells.
- T helper cells release cytokines and stimulate the selected B and T cells to divide (clonal expansion).
- T killer cells kill host cells that are infected with a virus. Some T cells become memory cells.

Once the correct B cell is found – again, clonal selection.

- It divides by mitosis to form a clone (collection of genetically identical cells) – clonal expansion.
- Some become memory cells.
- Some become plasma cells that release **antibodies** with a specific variable region complementary to the shape of the antigens originally presented.

✔ *Quick check 2*

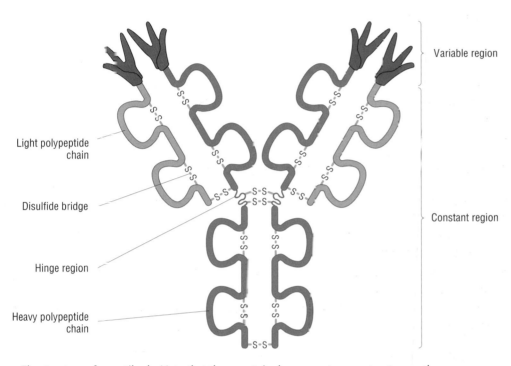

The structure of an antibody. Note that these proteins have a quaternary structure as they consist of more than one polypeptide chain.

Labels on diagram:
- Variable region
- Light polypeptide chain
- Disulfide bridge
- Hinge region
- Constant region
- Heavy polypeptide chain

Examiner tip

There is a different antibody for each antigen. What makes each antibody specific is the particular amino acid sequence at the variable region. It is different for each set of antibodies made by one clone of B lymphocytes. However, individual antibodies *cannot* change this region.

Hint

You may be asked how antibodies bind to antigens. They are both proteins so many hydrogen and ionic bonds form between their amino acids. Although these bonds are not as strong as covalent bonds, there are many of them.

Antibodies are released into the bloodstream. They may:

- coat bacteria so that phagocytes can ingest them
- cause viruses to clump together; this prevents them from entering cells
- coat bacteria preventing them from attaching to host cells
- cause lysis of foreign cells.

Special antibodies, IgE, are made in response to parasitic worm infections and these may also be involved in allergic reactions.

When the pathogen has been destroyed by the immune response, **memory cells** (B and T types) remain in the body for a very long time. This makes the person immune to that particular pathogen. If it invades the body again, the immune response is so quick that no symptoms are experienced.

✔ *Quick check 3*

QUICK CHECK QUESTIONS

1 Describe the non-specific inflammatory response.

2 Explain the different roles of T and B lymphocytes.

3 Explain why a person who suffered from flu last year may suffer from a different strain of flu this year.

Blood groups

Key definition

Agglutination is what happens when antibodies bond to several red blood cells, causing them to clump together.

When blood transfusions were first carried out some people survived and some died. Scientists later found that different people have different blood groups. Now people receiving a blood transfusion are given the correct type of blood.

The ABO system

There are many ways of categorising blood into groups but the **ABO system** is particularly important for blood transfusions.

On the cell surface membranes of erythrocytes are protein **antigens** (sometimes also called **agglutinogens**).

In the blood plasma are **antibodies**. The antibodies in a person should not be complementary to the antigens on their red blood cells, otherwise these cells would agglutinate (clump together). This is fatal as the blood supply to organs would be severely impaired. Damaged red cells would release haemoglobin that is toxic to tissues.

Blood group	Antigens on surface of red blood cells	Antibodies in plasma
A	A	Anti B
B	B	Anti A
AB	A and B	None
O	Neither A or B	Anti A and anti B

When considering blood transfusions we need to remember the following:
- The donated blood is diluted by the recipient's blood, so it is the antibodies in the recipient's blood that are going to cause clumping.
- Hence, people of blood group O can only receive group O blood as they have both types of antibodies in their plasma.
- People of blood group AB can receive any type of blood as they have no ABO antibodies.
- People of group A can receive group A and group O blood as group O blood has no ABO antigens on the red cells so there is nothing for the recipient's ABO antibodies to react with.
- People of group B can receive group B and group O blood.
- However, hospitals always try to match the ABO group of donor and recipient as full compatibility is safer and prevents other side effects.

We refer to blood types being **incompatible** if transfusions cannot be received.

Blood group	Can donate to	Can receive from
A	A and AB	A and O
B	B and AB	B and O
AB	AB	A, B, AB and O
O	A, B, AB and O	O

The compatibility of blood groups

Rhesus factor

Some people have another protein, called a D antigen (**rhesus factor**) on the surface of their red blood cells. If they have it they are described as rhesus positive. (It was found that this antigen, when injected into rhesus monkeys, caused agglutination.)

People without this antigen are described as rhesus negative.

You may recall that during antenatal care the medical staff find out the mother's blood group and whether she is rhesus positive or negative. If she is negative, they find out if her husband is rhesus positive. This is because her baby could be rhesus positive. That, in itself, is not a problem, but at birth her blood will be exposed to some of the baby's blood. She will then produce anti-rhesus antibodies that could later cross the placenta and harm a subsequent rhesus-positive fetus.

The baby when born has:
- an enlarged liver and spleen
- jaundice due to bilirubin released from destroyed red blood cells
- oedema
- difficulty breathing.

✔ *Quick check 2*

A rhesus-negative mother can be desensitised after the birth of her first rhesus-positive baby by being injected with antibodies against the rhesus-positive cells that are then destroyed before an immune response can be stimulated.

For blood transfusions the donor and recipient blood must be compatible. Rhesus-negative people cannot be given rhesus-positive blood.

QUICK CHECK QUESTIONS

1 Which of the following are compatible:
  (a) group A donor and group O recipient
  (b) group AB donor and group A recipient
  (c) rhesus-negative donor and rhesus-positive recipient?
Explain your answers.

2 Explain why a rhesus-positive mother carrying a rhesus-negative fetus is not a medical problem.

Tests and vaccines for TB, HIV and cervical cancer

Key words

- vaccine
- HIV
- human papilloma virus (HPV)
- cervical cancer
- TB

The World Health Organisation managed to eradicate smallpox by using vaccination. Wherever there was a case of smallpox they vaccinated anyone likely to come into contact with that person, thus breaking the chain of infection.

This method was successful for many reasons, including the following.

- The smallpox virus does not mutate so the vaccine was always effective.
- The freeze-dried vaccine did not degrade in hot climates.
- The inoculation was just under the skin so many volunteers could be trained to help administer it.
- Smallpox only infects humans so there were no animal reservoirs.

The biological problems with developing a vaccine against HIV

Hint

Read the question carefully – is it asking about HIV or HPV?

✔ *Quick check 1*

Not long after HIV was identified, in 1984, the US secretary of State for Health said that the discovery and identification of the virus would enable scientists to develop a vaccine to prevent AIDS. She expected that such a vaccine would be ready for trialling by 1986. Now, 20 years on, we still do not have an effective vaccine against HIV. This is because none of the classical methods of making a vaccine works with this virus. A vaccine's effectiveness depends upon the body's immune response. Vaccines either invoke the production of antibodies or of killer T cells. HIV disables or evades the immune system in the following ways:

- it invades and disables macrophages so there are fewer antigen-presenting cells
- it disables helper T cells, particularly memory helper T cells
- it also evades killer T cells
- it mutates within a host, so that any killer T cells that could recognise infected cells at the beginning of an infection, cannot do so later on. The diversity of antigens on HIV within a single person, after six years of being infected, is greater than the diversity of all the strains of flu virus, worldwide, within any given year. Therefore, making a vaccine for HIV would be like making a vaccine for hundreds or thousands of different viruses.

Ethical issues relating to the development of a vaccine for HPV

Human papilloma virus (HPV) causes genital warts and cervical cancer. This means that cervical cancer is a sexually transmitted disease.

There is now a vaccine against this virus.

- It needs to be given to girls before they become sexually active.
- This means that girls aged 12 years should be vaccinated.
- Some people object as they fear it will encourage promiscuity and encourage girls to become sexually active before they reach age 16 years.
- However, if a vaccine is available that can prevent cancer, is it ethical to not give it? Should males also be vaccinated as they can become infected with this virus and then infect their partners?
- Not all cases of cervical cancer are attributable to HPV – will the vaccine mean that fewer people will attend for smear tests? Will this mean cases that do occur will go unrecognised until it is too late for treatment?

✔ *Quick check 2*

How individuals can be tested for HIV and TB infections

Disease	Test
HIV	**PCR (polymerase chain reaction):** [see page 127] • HIV is a retrovirus • It contains RNA which it inserts into the host T helper lymphocyte cells • Using its enzyme, reverse transcriptase, it makes copies of DNA from the RNA and incorporates it into the host cell • This test extracts and copies sections of the DNA from these cells, to amplify it and produce enough to analyse, to see if viral DNA is present This test is usually carried out on newborn babies whose mothers are HIV+
	HIV antibody test: this is used on adults • People infected with HIV will, after about 3 months, produce antibodies against the virus • The test uses another antibody that binds to the HIV antibody • The test antibody has an enzyme attached to it • If the HIV antibodies are present in the blood sample, the test antibodies and enzymes become attached to it Add blood sample HIV antibody HIV antigen Enzyme Second antibody • When a chemical is added, the enzyme causes it to change colour, showing a positive result The test cannot be carried out immediately after suspected infection and is not infallible, so could produce false positives or false negatives; people may need counselling or advice when they receive the results of the test Screening is carried out of donated blood for the blood transfusion service
TB	The **Mantoux test** is used to find out if people have been exposed to TB bacteria; this has replaced the Heaf test previously used in the UK • A small amount of serum containing TB antigens is inoculated under the skin of the forearm • The extent of the swelling after 48–72 h is measured • Those with a positive result can be investigated to see if they have TB; those with a negative result may be given the BCG vaccination

✓ *Quick check 3*

QUICK CHECK QUESTIONS

1 Explain why it is unlikely that an effective vaccine against HIV will be produced.

2 Discuss the ethics of vaccinating 12-year-old girls against HPV.

3 Describe the test carried out to see if a person has been exposed to TB.

UNIT 2 Epidemiology

Key words

- epidemiology
- chronic disease
- morbidity
- mortality
- incidence
- prevalence
- notifiable disease

Key definition

Epidemiology is the study of the factors affecting the health and illness of populations. Research on the occurrence and distribution of diseases in populations is used to inform policy making about the best treatment for particular diseases or the best way to prevent them.

Key definition

Chronic diseases are diseases of slow onset that gradually get worse. There is no cure but there may be treatment. However, most are preventable by adopting an appropriate lifestyle earlier in life. They include, type 2 diabetes, coronary heart disease (CHD) and arthritis.

Epidemiologists study the factors that affect the health of populations. The data generated by epidemiological studies are used:

- to identify risk factors for specific diseases
- to see if a disease is endemic, epidemic or pandemic
- to determine **morbidity** and **mortality** rates for diseases
- to determine **incidence** and **prevalence** of diseases
- to provide data for evidence-based medicine
- to identify countries at risk and which parts of the population are at risk
- to target vaccination programmes
- to target education programmes at the most at-risk groups of people
- to target screening programmes
- to target research to find cures or better treatments for diseases
- to compare different primary care trusts or hospitals
- to evaluate the success of control programmes.

It is a multidisciplinary subject and uses biology, sociology, psychology, maths and statistics.

Most diseases, even acute infectious ones caused by an infecting agent (pathogen), are affected by other factors. For example, nutrition status, living conditions (e.g. overcrowding, damp conditions) and stress levels will all affect whether or not the person develops the disease, and how well or badly their own immune system deals with it.

Quick check 1

The diseases of chronicity (chronic diseases) that affect modern Western society, such as obesity, diabetes, heart disease, stroke and arthritis, are multifactorial. Hence, there is no simple 'one cause – one effect'. In addition, correlation does not necessarily mean cause, although it means that the factor(s) in question *may* contribute to the cause of the disease.

Epidemiologists use data on the following:

Morbidity rate = number of cases of a particular disease in a single year in a specified population unit, such as x cases per 1000 people. It may also be calculated on the basis of age groups, sex, occupation or other population unit such as smokers. Bear in mind that some people suffer from more than one disease/illness.

Mortality rate = the death rate per unit of population, such as deaths per 1000, per 10 000 or per 100 000 people, per year.

Incidence = the number of new cases of a disease in a particular period.

Incidence rate = the number of new cases in a specific population in a certain time period – this is one example of a morbidity rate.

Prevalence = the total number of cases, both old and new, of a disease, present during a particular time period.

Prevalence rate = the total number of cases of a disease, per unit of population, per time period – this is another example of a morbidity rate.

Notifiable diseases

Doctors in England and Wales must report to their local authority on suspected cases of certain infectious diseases, such as mumps, rabies, rubella, polio, anthrax, cholera, food poisoning, malaria, measles, meningitis, tetanus, TB, viral hepatitis and whooping cough. They also report births and deaths, and violations of public health.

Each week, the local authorities inform the Health Protection Agency (HPA) centre for infections, which then collates and analyses these data, and publishes information on national and local trends.

Examiner tip

In calculating rates, it is usual to give the answer to the nearest whole number: you do not get 0.5 of a person.

✓Quick check 2

✓Quick check 3

(a)
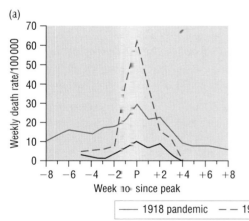
(b)
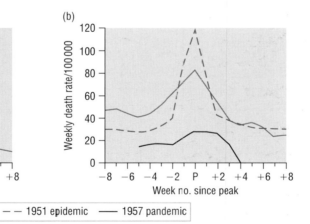

—— 1918 pandemic – – 1951 epidemic —— 1957 pandemic

Effects of three different influenza pandemics in Liverpool

QUICK CHECK QUESTIONS

1 The incidence of many infectious diseases started to fall in the UK, long before vaccinations were introduced. Suggest what factors caused this fall in incidence.

2 Explain why, during a particular year, the prevalence of a disease could be high but the incidence could be low.

3 Suggest why doctors have to notify their local authorities about new suspected cases of certain infectious diseases.

The importance and global impact of TB and HIV

Key words

- HIV
- infectious disease
- pandemic
- biological factors
- social and economic factors
- ethical factors
- TB
- control
- prevention
- migration

| Western and Central Europe **720 000** | Eastern Europe and Central Asia **1.6 million** | East Asia **870 000** |

North America **1.2 million**

Caribbean **300 000**

North Africa and Middle East **510 000**

South and South-East Asia **7.4 million**

Latin America **1.8 million**

Sub-Saharan Africa **25.8 million**

Oceania **74 000**

Estimated numbers of people infected with HIV

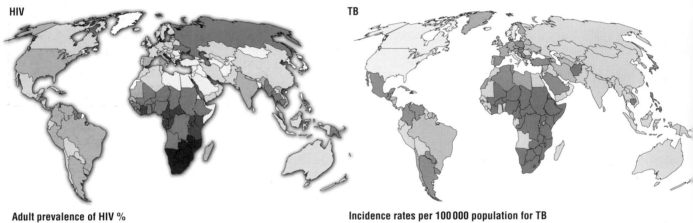

HIV

TB

Adult prevalence of HIV %

| ■ 15.0 – 34.0 | ■ 1.0 – < 5.0 | 0.1 – < 0.5 |
| ■ 5.0 – < 15.0 | 0.5 – < 1.0 | <0.1 |

Adult prevalence of HIV

Incidence rates per 100 000 population for TB

| Less than 10 | ■ 25 to 49 | 100 to 299 | ■ No estimat |
| ■ 10 to 24 | 50 to 99 | ■ 300 or more | |

Incidence rates, per 100 000 population, for TB

HIV is a recent worldwide **infectious disease** that is spreading in **pandemic** proportions. By the end of 2005, there were approximately 45 million people infected with HIV/AIDS and nearly 30 million people had died from HIV/AIDS-related illnesses. Over half the cases of HIV/AIDS are in sub-Saharan Africa. During 2007–2008 the number of cases in Russia, other Eastern European countries and China was increasing rapidly.

✓ *Quick check 1*

There are many reasons why the disease is spreading on such a large scale.

Biological factors	Social and economic factors	Ethical factors
• Virus mutates and changes its antigens so there is no vaccine • It can be sexually transmitted and it is difficult to get people to change their (sexual) behaviour • There are no obvious symptoms so people often do not know they are infected	• Many countries with a high prevalence of HIV are less economically developed countries (LEDCs). Their governments have less money to spend on preventive strategies, such as screening blood for transfusions, supplying sterile needles for injections and giving free condoms to people • Some governments, e.g. in South Africa, give out incorrect advice • If people are poorly educated they do not know how to protect themselves or their unborn children from the disease • Where there is poverty, more women are forced to work in the sex industry and they can earn more money by having unprotected sex • Men have to move to cities to find work, and when there they may visit prostitutes, become infected and then infect their wives • HIV affects many young adults and this leads to loss of an important section of the workforce	• In areas of civil unrest, rape and deliberate infection with HIV have been used as weapons of war • Women in LEDCs have low status and men are unwilling to use condoms • People with HIV are often stigmatised and so people may be unwilling to be tested

TB is also a worldwide disease. It has infected humans for hundreds of years but during the nineteenth century and early twentieth century, improved living conditions, antibiotics and vaccination appeared to have drastically reduced the incidence of the disease. However, it has re-emerged. The HIV pandemic has also led to an increase in numbers of TB cases, as TB is an opportunistic infection and affects people with weakened immune systems. In 2005 about 8.8 million new cases were recorded and about 1.6 million people died of TB.

TB has been hard to **control** due to many factors.

✔ *Quick check 2*

Biological factors	Social and economic factors	Ethical factors
• Some strains of the bacterium are resistant to many antibiotics so treatment involves taking many antibiotics for a long period. Many people do not adhere to the treatment regime • Vaccine is not very effective, particularly in LEDCs where children are less well nourished	TB spreads by droplets and readily where people: • live in sub-standard housing • live in overcrowded houses • have a poor diet • drink unpasteurised milk People may not be educated and therefore do not understand how TB spreads or how to reduce risk of infection There may be lack of available transport for people to access healthcare centres	Some countries cannot afford to provide good healthcare and cannot isolate patients Many homeless people suffer from TB. They become infected when they spend nights in shelters, in large dormitories, where beds are close together and the bacteria spread by droplet infection. Being homeless means their general health status and nutrition may be poor and their immune systems are weak. It is also harder to access them and give effective medical treatment or to follow up and check their progress

Future pandemics

Epidemiological data can tell us when a pandemic is on the way. Flu nearly always spreads from east to west, so if we know the strain of virus that is going to infect us we can administer vaccines to the most vulnerable people to **prevent** it.

The following are potential reasons why a future pandemic may occur.

- Containment – free trade, freedom of movement and tourism make it difficult to contain an outbreak. Where animals are a secondary host **migration** of the animals could spread the pathogen.
- Lack of herd immunity – so most people who encounter the new pathogen will be susceptible.
- Lack of vaccine stocks – with a new strain, existing stocks of vaccine may be ineffective.
- Healthcare workers will be vulnerable and containment facilities will be overstretched.

QUICK CHECK QUESTIONS

1 Discuss the social biological and ethical reasons why HIV/AIDS is such a globally important disease.

2 Discuss the social, biological and ethical reasons why TB is such a globally important infectious disease.

Coronary heart disease (CHD) –1

Key words

- non-infectious disease
- infectious disease
- CHD
- angina pectoris
- myocardial infarction/heart attack
- cardiac arrest
- atherosclerosis
- first aid
- cardiopulmonary resuscitation (CPR)
- defibrillators
- aspirin

✔ *Quick check 1*

Key definition

A **non-infectious** disease is one that is not caused by a pathogen and is not transmitted from person to person. CHD is non-infectious.

An **infectious disease** is caused by a pathogen invading the body. An example is measles.

CHD is a result of having diseased coronary arteries. This may lead to ischaemic heart disease, where blood flow to the heart muscle (myocardium) is restricted. This may take the form of **angina pectoris** or a **heart attack** (**myocardial infarction**). Following a heart attack, **cardiac arrest** may occur.

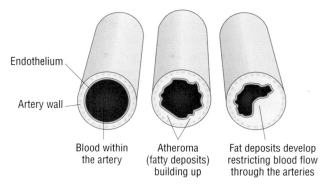

Endothelium

Artery wall

| Blood within the artery | Atheroma (fatty deposits) building up | Fat deposits develop restricting blood flow through the arteries |

How atherosclerosis develops

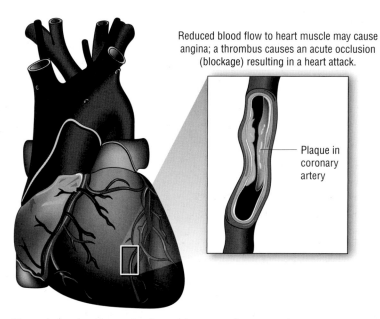

Reduced blood flow to heart muscle may cause angina; a thrombus causes an acute occlusion (blockage) resulting in a heart attack.

Plaque in coronary artery

How atherosclerosis may lead to angina or to a heart attack

Atherosclerosis

- Fatty deposits, called atheromatous plaques, build up in the wall of arteries, beneath the endothelium.
- Fibrous tissue and calcium deposits invade these plaques.

Where coronary arteries are narrowed, angina results.

If the plaque ruptures and causes a blood clot in the coronary arteries, then a heart attack occurs.

Angina pectoris	Heart attack
• The plaque narrows arteries and reduces blood flow, and hence supply of oxygen and fatty acids, through them • Heart muscle obtains more than 75% of its ATP by respiring fatty acids • There may be enough oxygen to meet normal needs, but if the person starts to exert, the heart muscle cannot respire fatty acids, which must be aerobically respired • Anaerobic respiration of glucose releases only small amounts of ATP, and with an occluded coronary artery not enough glucose can be supplied • Histamines, proteolytic enzymes and other cell signals are released and these stimulate nerve endings, causing the pain of **angina pectoris** • The (referred) pain is beneath the upper sternum, down the left arm, in the left shoulder, neck and jaw; the person also feels as if they are suffocating	• The membrane covering the plaque may rupture and red blood cells stick to the plaque, forming a thrombus • The coronary artery is blocked • The muscle beyond the blockage does not receive any blood (or oxygen/fatty acids) and is infarcted. The overall process is a **myocardial infarction** or **heart attack** • Heart muscle cells in the area die and the tissue is invaded by fibrous scar tissue Symptoms are: • severe chest pain, as if a tight band is around the chest • breathlessness, nausea and sweating • irregular pulse • ashen cold skin and blue lips **First aid** help is to make the person comfortable and phone for an ambulance If treated with drugs to dissolve the thrombus, or if angioplasty and stent insertion are carried out, the person recovers The heart attack may lead to death by reduced cardiac output leading to cardiac shock: • reduced blood flow to kidneys and hence toxic waste cannot be excreted • reduced return of blood to the heart causing damage to lung capillaries • heart rupture • cardiac arrest **Cardiac arrest** • The ventricles fibrillate • The muscle cells beat in an uncoordinated way so the ventricles do not contract properly • This may occur within 10 min of the infarction **Immediate cardiopulmonary resuscitation (CPR)** must be given, whilst waiting for an ambulance, otherwise brain, heart, lung and kidney damage, and death will occur

CPR involves rescue breaths and chest compressions.
For adults, give 30 compressions with the heel of the hand in the centre of the chest, pressing down 4–5 cm at a rate of 100 per minute. Give two rescue breaths by pinching the nose, sealing lips round casualty's mouth and blowing until the chest rises. Continue until emergency help arrives or the casualty breathes normally.

✓ *Quick check 2*

In hospitals, **defibrillators** are used. These pass a large amount of electricity through the heart, which stops the heart. When it restarts, it should be in a synchronised fashion.

Medication for angina

People suffering from angina and at risk of heart attack (or stroke which is ischaemia in the brain) may take **aspirin** each day to prevent formation of blood clots in vessels.

✓ *Quick check 3*

Other medication includes:
• nitroglycerine to dilate the blood vessels
• beta blockers to block nerve impulses that normally cause heart rate to increase.

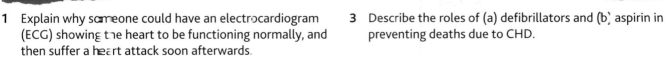

QUICK CHECK QUESTIONS

1 Explain why someone could have an electrocardiogram (ECG) showing the heart to be functioning normally, and then suffer a heart attack soon afterwards.

2 Explain the difference between a heart attack and cardiac arrest.

3 Describe the roles of (a) defibrillators and (b) aspirin in preventing deaths due to CHD.

UNIT 2

Coronary heart disease (CHD) –2

Key words

- genetic factors
- social factors
- behavioural factors
- distribution
- risk
- diet
- blood pressure
- exercise
- smoking
- BMI
- waist/hip ratio
- coronary bypass surgery
- angioplasty
- heart transplant
- NICE (National Institute for Health and Clinical Excellence)

Global distribution of CHD

Numbers of early deaths from CHD vary throughout the world (see graph below). Epidemiologists use these data to find out the risk factors for this disease.

Risk factors for CHD

There are three main determinants of health.

- **Genetic factors** – some people have a higher genetic predisposition to heart disease.
- **Environmental or social factors** – this may relate to exposure to certain substances or to stress caused by where you live.
- **Behavioural factors** – including the lifestyle choices you make.

The main **risk** factors are:

- **dietary**
  - ○ high fat
  - ○ high saturated fat
  - ○ high salt
  - ○ low on fresh fruit and vegetables
  - ○ low on oily fish
- having diabetes
- high alcohol consumption
- **smoking**
- lack of exerise

- age – the arterial endothelium becomes rougher and fat deposits form more readily
- gender – until menopause women's risk is lower, but after age 50 it is equal or greater
- hypertension (high resting **blood pressure**) – in some cases this is genetic and can be treated with drugs; in many cases a high salt diet is the main factor or arteriosclerosis (loss of elasticity in the arteries with age)
- maternal nutrition – what your mother ate whilst you were in the uterus has a long-lasting effect on your health; if she ate a diet low in protein you are more at risk of early CHD
- being overweight or obese.

✓ *Quick check 1*

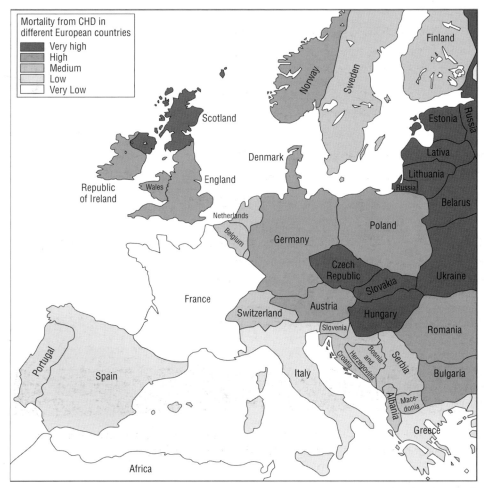

Mortality from CHD in different European countries

These factors may be different for different parts of the world and so influence the distribution of CHD. Whereas in less economically developed countries (LEDCs) richer people may be overweight, in more economically developed countries (MEDCs) it is poorer people who are more likely to be overweight.

CHD is regarded as a disease of affluent countries, but within those countries it is the poorer sections of society who suffer most. This may be due to a number of factors, such as:

- fresh fruit is expensive
- lean meat is more expensive than fatty meat
- cheaper food and processed meals contain a lot of saturated fat and salt
- people who are less well off are more stressed as they have less control over their lives; they may therefore smoke and drink more alcohol.

✓ *Quick check 2*

BMI and waist measurement

Having a **body mass index (BMI)** of more than 25 signifies overweight and over 30 signifies obese. The BMI can be calculated as:

$$BMI = \frac{mass\ in\ kg}{(height\ in\ m)^2}$$

However, **waist measurement** is a better indicator of high risk as having a lot of body fat around your middle (being apple shaped) is more dangerous than having weight on the hips (being pear shaped). If the waist measurement of a man is above 39 inches, and of a woman above 36 inches, the risk of CHD is greatly increased.

✓ *Quick check 3*

Main medical treatments for CHD

Coronary bypass surgery – a piece of vein is taken from the arm or leg, or a piece of artery from a mammary gland, and attached to the coronary artery beyond the blockage so that blood can flow through it and bypass the blocked portion of the artery.

Angioplasty – a balloon and mesh stent are inserted into the coronary artery. The balloon is inflated. This squashes the plaque, widening the lumen of the artery. The balloon is withdrawn and the stent left in place.

Heart transplant – where donor and recipient are matched for tissue type. The recipient has to take immunosuppressant drugs afterwards to prevent rejection.

The cost of treating cardiovascular diseases is high as many people now live longer and may develop CHD.

NICE (National Institute for Health and Clinical Excellence) produces guidance to the NHS about clinical practices and medicines and treatments that are deemed cost effective. It also gives guidance to NHS and local authority health promotion professionals on the promotion of good health and prevention of ill health.

Prevention is always better than treatment, both for individuals and for the health service, but it is difficult to get people to change their behaviour.

QUICK CHECK QUESTIONS

1   What advice would you give to someone who is overweight, smokes and drinks 20 units of alcohol every weekend?

2   Explain why it is incorrect to describe CHD as a disease of affluence.

3   Suggest why waist measurement is a better indicator than BMI for risk of heart disease.

Lung disease

Key words

- acute
- chronic
- lung cancer
- asthma
- COPD
- beta$_2$ agonists
- steroids

Key definition

An **acute** disease is one that has sudden onset and can be cured. Examples include measles and influenza.

A **chronic** disease is one that has gradual onset and can be treated but not cured; however, it may be prevented. Examples include most types of cancer, chronic obstructive pulmonary disease and coronary heart disease (CHD).

Short-term effects of smoking on the respiratory system

Component of tobacco smoke	Short-term effect on respiratory system
Carbon monoxide	• Combines irreversibly with haemoglobin, forming carboxyhaemoglobin. This reduces the oxygen-carrying capacity of the blood (not all the haemoglobin in all red cells will be affected), so tissues receive less oxygen for aerobic respiration • If a pregnant woman smokes her fetus receives less oxygen and this reduces its growth
Tar	• Causes goblet cells to enlarge and to produce more mucus • Paralyses or damages cilia so they cannot waft and move the mucus. This increases the likelihood of lung infections

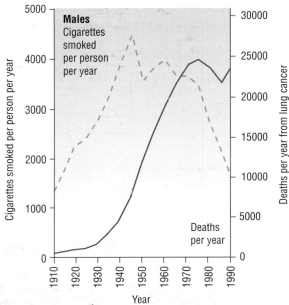
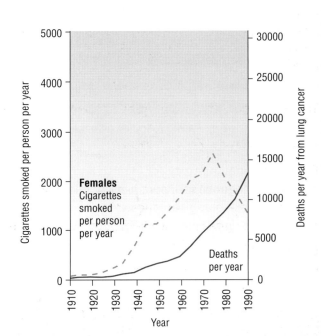

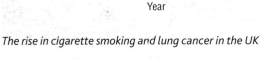
The rise in cigarette smoking and lung cancer in the UK

Long-term effects of smoking on the respiratory system

Condition	Description of development	Symptoms
Lung cancer	Tar contains carcinogens such as benzopyrene • This causes mutations in genes that control cell division, in cells of the bronchial epithelium • It also inactivates the p53 protein • Cells divide uncontrollably and form a tumour	Lung cancer takes about 30 years to develop, so a smoker appears healthy for a long time, but then they: • feel breathless • wheeze • have a persistent cough • cough up blood • experience voice changes ✓*Quick check 1*
Emphysema	Increased lung infections, due to excess mucus that can't be removed, lead to: • macrophages secreting an enzyme, elastase, that breaks down elastin in alveoli walls • Alveoli can't stretch and recoil and they burst • This reduces surface area for gaseous exchange ✓*Quick check 2*	• Breathlessness • Barrel chest • Unable to carry out any exercise • May eventually be unable even to walk upstairs • May need to be linked to oxygen cylinder
Chronic bronchitis	Increased production of mucus in trachea and bronchi This leads to infections of respiratory tract and scar tissue forming in the airways	• Cough with phlegm • Wheezing • Breathlessness on exertion • Blue lips • Can lead to (right ventricle) heart failure
Asthma	Chronic inflammation of the airways or a sudden spasm of the smooth muscle of the airways Some attacks may be the result of allergens There is a high incidence in some isolated populations due to a genetic defect	• Breathlessness – able to inspire but can't breathe out very well • Wheezing • Prolonged expiration and coughing • May lead to an enlarged 'barrel' chest • May be fatal
Chronic obstructive pulmonary disease (COPD)	Combination of asthma/emphysema/chronic bronchitis Progressive and irreversible	• Reduced lung capacity • Difficulty breathing in and out deeply • Cough • Unable to exert/exercise

These conditions may result from active or passive smoking.

Medications for asthma

Relievers:
- are **beta₂ agonists** that are bronchodilators, which relax smooth muscle in airways
- may be short-acting or long-acting
- are usually inhaled as needed but sometimes taken orally.

Controllers:
- are **steroids** (corticosteroids) that reduce inflammation
- are usually inhaled but may be taken orally.

✓*Quick check 3*

QUICK CHECK QUESTIONS

1 Smoking became very popular with men in the UK just after World War 1. Explain why the huge increase in lung cancer cases in men in the UK was not seen by epidemiologists until the late 1940s.

2 Explain why people with emphysema cannot walk upstairs.

3 Suggest when an asthmatic would use (a) a beta₂ agonist inhaler and (b) a corticosteroid inhaler.

Types and distribution of diabetes

Key words

Key words

- type 1 (insulin-dependent) diabetes
- type 2 (non insulin-dependent) diabetes
- diet
- distribution within populations
- economic development

Key definition

Diabetes mellitus, or sugar diabetes, is a disorder of carbohydrate, fat and protein metabolism that is the result of a deficiency of insulin secretion by beta cells in the pancreas, or resistance of target cells to insulin. It is usually referred to as 'diabetes'.

There is a pandemic of diabetes (type 2), mainly associated with lack of exercise and obesity. This is beginning to affect many countries as they **develop economically**. There is also a genetic component to type 2 diabetes so when some groups of people migrate to other areas, this may alter the incidence of type 2 diabetes in that part of the world. The cost to a health service for treating type 2 diabetes is enormous and will put a huge strain on it. The cost to individuals is also high as it reduces quality of life and shortens lives.

✔ *Quick check 1*

Types of diabetes

	Type 1	Type 2	Gestational
Cause	Autoimmunity • The body's immune system destroys insulin-producing beta cells in the islets of Langerhans in the pancreas • This may be after an infection with a type of virus that has similar antigens to a chemical on the surface of beta cells	Target cells are resistant to insulin. May be a mutation in gene coding for insulin receptors Beta cells may also become 'worn out' as they have to produce more insulin, so type 2 can become insulin dependent Associated with: • being obese • eating a high fat and sugar diet • being sedentary • ageing • genetic predisposition	Glucose intolerance develops during pregnancy
Onset	Suddenly In children	• Usually in mid life, from age 40 years • However, it is now occurring in younger people, including adolescents and children • This correlates with the huge increase in obesity in young people	In some women during pregnancy Usually during third trimester
Characteristics	If untreated leads to vomiting, mental confusion and diabetic coma caused by ketoacidosis – a build up of ketones in the blood, lowering its pH, as the body metabolises fats instead of carbohydrate	• Slow onset • Progressive • Weight loss • More urine produced • Hunger • Raised blood glucose level • Much glucose in urine • Atherosclerosis develops; therefore increased risk of CHD and stokes • Blindness • Kidney failure • Chronic infections • Wounds fail to heal and ulcers develop • Circulatory problems leading to gangrene and loss of feet or lower legs • Inflammation of nerve endings (neuropathy) • Reduces life expectancy by 10 years • Ketoacidosis and coma ✔ *Quick check 2*	

Treatment	Emergency treatment for coma is insulin injections Treatment for low blood sugar is glucose or sugary food Management: • daily insulin injections at mealtimes • regular meals • balanced diet • exercise	May be managed by: • balanced **diet** • increased exercise ✓ *Quick check 3* • drugs (tablets), such as metformin, to lower the blood glucose level; drugs are used with exercise and diet changes, NOT instead of them • paracetamol or ibuprofen for neuropathy Emergency treatment for coma is insulin injection Treatment for low blood sugar is glucose or sugary food May eventually need daily insulin injections	
Distribution	About 0.3% of UK population An estimated 4.9 million cases globally Rising most rapidly in Europe Finland has highest prevalence with 40 per 100 000 population, reason unknown	About 2.7% of UK population but a further 1 million may have it and not know it Greater prevalence amongst Asian and African/Caribbean groups Worldwide, 200 million cases and increasing	Occurs in about 1%–14% of pregnancies About half of these will develop diabetes mellitus within 15 years of the pregnancy

There are other causes of diabetes:

• damage to the pancreas
• associated with conditions such as Down's syndrome, Turner's syndrome and Klinefelter's syndrome.

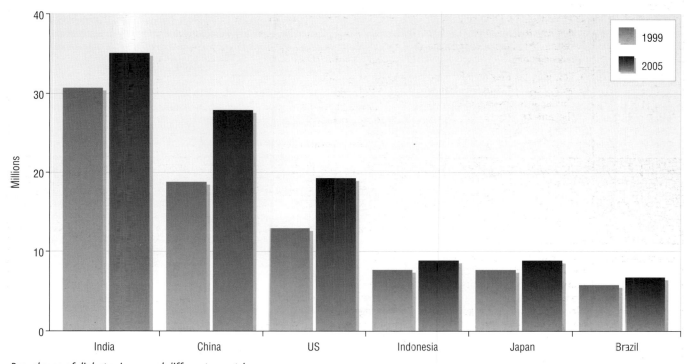

Prevalence of diabetes in several different countries

QUICK CHECK QUESTIONS

1 Explain why the huge increase in type 2 diabetes mellitus is of concern to governments and health services.

2 Describe the problems associated with type 2 diabetes mellitus.

3 What lifestyle behaviours should people adopt and encourage their children to adopt, to reduce the risk of developing type 2 diabetes?

Diagnosing type 2 diabetes and monitoring blood glucose levels

Diagnosing type 2 diabetes

A simple test could be to check for the presence of glucose in urine. However, in healthy people there are often small amounts of glucose in urine and the 'dip stick' test strips are not very sensitive or accurate in showing how much glucose is present.

A better test is to test the blood for glucose levels. In early stages of type 2 diabetes, insulin is present. However, as the target cells do not respond to it and take up more glucose from the blood, so the blood glucose level remains high after eating.

Test	How it is carried out	How results are interpreted
Fasting glucose test	• Carried out early in the morning, after the patient has fasted (not eaten anything) for at least 8 hours • Patient given a drink containing 75 g glucose • Blood taken after 2 hours and tested for glucose level	• If the level of glucose in the blood is higher than 6 mmol dm^{-3} then there is impaired glucose tolerance • If the level is above 7 mmol dm^{-3} then the patient may have diabetes • Further tests will need to be done to make sure
Random glucose test	Patient given a glucose drink at any time	Level of glucose in blood measured and compared with published data
Glucose tolerance test	• Patient fasts for 8 hours before test • Blood glucose level measured • Patient given a glucose drink containing 75 g glucose • Blood glucose level measured every 30 minutes for 3 hours	Blood levels of glucose above 11 mmol dm^{-3} after 2 hours indicate probable diabetes

The glucose tolerance test is not considered very reliable.

Further tests measuring the amounts of insulin in the blood may be carried out to verify the diagnosis.

Monitoring blood glucose levels

As mentioned earlier, **blood glucose level** is a more accurate way of monitoring the effective management of diabetes, as many healthy people have small amounts of glucose in the urine.

A person with diabetes should try to make sure that their blood glucose concentration is:
- 4–7 mmol dm⁻³ before meals
- Less than 10 mmol dm⁻³ 90 min after a meal
- About 8 mmol dm⁻³ at bedtime.

A diabetic person can use a **biosensor** to measure the glucose level in a small drop of blood. Biosensors are available cheaply from pharmacists.

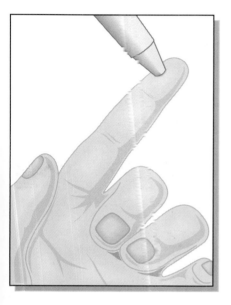
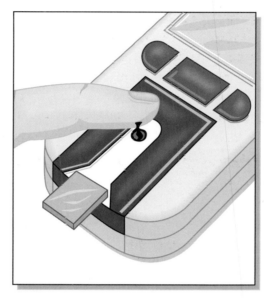

Using a biosensor to test the blood glucose concentration

1 Place a test strip in the biosensor.
2 Clean the fingertip with alcohol and use a lancet to make a small prick in the fingertip.
3 Squeeze a drop of blood onto the test strip.
4 After about 30 seconds note the reading.

The biosensor works as follows.
- In the biosensor strip is an enzyme that converts glucose to gluconolactone.
- A small electric current is produced.
- An electrode in the test strip converts this current to a numerical value.
- This is displayed digitally on a screen to show the blood glucose concentration.

The biosensor has a memory so results over a period of time can be kept and compared.

> **Hint**
>
> mmol dm⁻³ is a measure of concentration.
> A concentration of 1 mol dm⁻³ is 180 g glucose in 1 dm³ (1 litre) of solution. So 1 mmol dm⁻¹ – is 0.18 g glucose in 1 dm³.
>
> Normal blood glucose levels are between 0.75 g and 1.0 g per dm³, or 4–6 mmol dm⁻³.

✔ *Quick check 3*

QUICK CHECK QUESTIONS

1 Explain why testing the blood glucose level is a more reliable test for diabetes than testing for glucose in urine.

2 Describe how the fasting blood glucose test is carried out.

3 Describe and explain how a biosensor can be used to test blood glucose level in a person with diabetes.

End-of-unit questions

1 Mary is 16 years old and has been feeling unwell. She went to the doctor and was told that she had type 2 diabetes. Mary thought that type 2 diabetes was a disease of older people.

   Mary often goes out with her friends and often rounds off the evening with a burger and chips and a bottle of fizzy drink. She has been gaining too much weight over the last few years.

   The doctor gave Mary a glucose test monitor to test her blood glucose concentration every morning before breakfast.

   (a) Explain why the doctor wanted Mary to:

   (i) measure her blood glucose concentration

   (ii) measure it every morning before breakfast.   (5)

   The doctor explained to Mary that it was essential she changed her diet.

   (b) State the changes that Mary should make to her diet.   (3)

2 There are a number of diseases where the immune system does not function as well as it should.

   These immunodeficiency diseases are either **inherited at birth** or **acquired** during an individual's lifetime.

   (a) Explain how immunodeficiency might be acquired.   (2)

   (b) One inherited type of immunodeficiency disease causes an absence of B lymphocytes and plasma cells.

   (i) Explain the connection between B lymphocytes and plasma cells.   (2)

   (ii) State the type of organism that would *not* be destroyed by an immune system without B lymphocytes.   (1)

   (iii) This immunodeficiency disease does not produce any obvious symptoms and therefore is rarely diagnosed until the individual is between two months and two years of age.

   Suggest why symptoms may not occur *until this age*.   (3)

   Severe combined immune deficiency (SCID) is a severe form of inherited immunodeficiency disease where both B *and* T lymphocytes do not form. Symptoms develop during the first few months of life and will result in death during the first two years, unless the infant is kept in a sterile environment.

   (c) Discuss the **ethical** issues affecting the families in which severe inherited diseases occur.   (4)

3 Humans start life as a single fertilised ovum. This cell has the potential to divide by mitosis to make all the cells of a new individual.

   (a) Describe the behaviour of chromosomes during anaphase of mitosis.   (3)

   (b) (i) Explain how the cell cycle produces genetically identical daughter cells. (3)

   (ii) State *two* roles of mitosis in a living organism.   (2)

   The changes which occur during the cell cycle are of great interest to professionals specialising in the study of cancer.

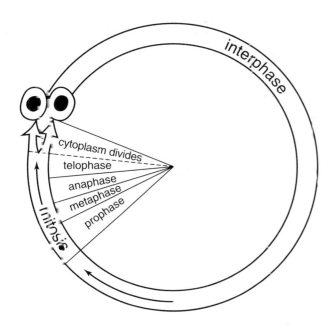

Figure 3.1 shows the main stages of the cell cycle.

(c) Identify the stage at which each of the following occurs:
   (i) DNA is replicated (1)
   (ii) the spindle forms (1)
   (iii) mitochondria within the cell replicate. (1)

(d) The anti-cancer drug vinblastine prevents spindle formation.
Suggest how this drug may slow down or prevent the development of a
tumour. (2)

Meiosis is another form of nuclear division.

(e) Explain the importance of **meiosis** in the human life cycle. (3)

4 Prenatal diagnosis is a widely used approach for the detection of sickle cell anaemia.
Chorionic villus sampling (CVS) allows prenatal diagnosis as early as 8 weeks.

(a) (i) Describe how CVS is carried out. (3)
   (ii) State *one* disadvantage of using CVS for the prenatal diagnosis of
conditions such as sickle cell anaemia. (1)
   (iii) Describe *two* ethical issues related to the prenatal diagnosis of sickle
cell anaemia. (2)

One method of measuring fetal growth is to measure, from an ultrasound scan,
the length of the back from the crown of the fetus to its rump.
Table 4.1 shows the changes in mean fetal crown–rump length during gestation.

Gestational age of fetus (weeks)	Mean crown to rump length (mm)
12	57
16	112
20	160
24	203
28	242
32	277
36	313
40	350

(b) Use the data in the table to calculate the percentage increase in mean crown–rump length from 36 to 40 weeks.
Show your working. (2)

A karyotype may be produced using chromosomes from cells collected during CVS or during an amniocentesis.
(c) Describe how the karyotype of a fetus with either of the following genetic disorders would differ from a normal karyotype:
  (i) a fetus with Turner syndrome
  (ii) a fetus with Klinefelter syndrome. (2)
(d) Suggest *one* advantage and *one* disadvantage of using an amniocentesis test rather than CVS. (2)

5 Growth patterns of children are used by health specialists to help diagnose problems of development at an early stage.

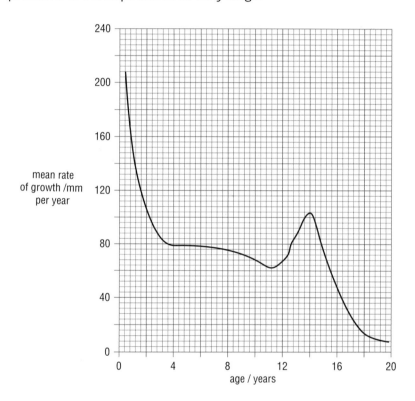

Figure 5.1 shows the mean rate of growth of humans from birth to 20 years.
(a) Describe and suggest reasons for the trends shown by the data in Figure 5.1. (4)
(b) Explain how you could determine the mean growth rate of a group of children over a 12-month period. (2)
(c) Name *two* nutrients needed by infants and state their role in maintaining healthy growth. (4)

6 A person can become immune to a specific infectious disease in several different ways.

Table 6.1 gives information about different types of immunity.

(a) Complete Table 6.1 by filling in the information for each type of immunity. The first one has been done for you.

Type of immunity	Gives immediate protection	Gives long-lasting protection
Passive natural	yes	no
Active natural		
Passive artificial		
Active artificial		

(3)

(b) During an immune response, antibodies are produced.

(i) Name the specific type of cell that produces antibodies. (1)

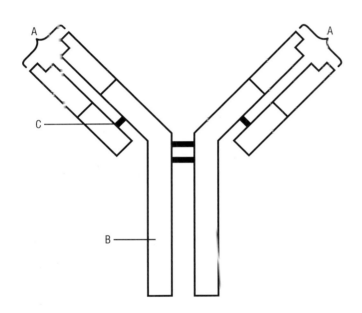

Figure 6.1 shows the structure of an antibody molecule.

(ii) Name **A**, **B** and **C**. (3)

(iii) Explain why one type of antibody will only bind to one antigen. (2)

(c) Describe how vaccination causes the development of immunity to an infectious disease.

In this question, one mark is available for the quality of spelling, punctuation and grammar. (7)

Quality of written communication (QWC: 1)

7 In the future, therapeutic cloning may produce new tissues and organs for people who are seriously ill.

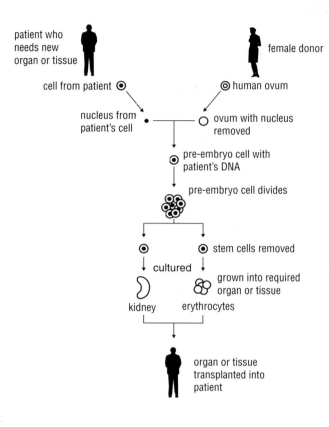

patient who needs new organ or tissue

cell from patient ⊙

nucleus from patient's cell •

female donor

⊙ human ovum

○ ovum with nucleus removed

⊙ pre-embryo cell with patient's DNA

pre-embryo cell divides

⊙ 　 ⊙ stem cells removed

cultured 　 grown into required organ or tissue

kidney 　 erythrocytes

organ or tissue transplanted into patient

Figure 7.1 shows stages in producing tissues and organs by therapeutic cloning.

(a) What is meant by *cloning*? (1)

(b) Using Figure 7.1:

　(i) explain why stem cells are used in cloning (2)

　(ii) outline how stem cells are cultured (2)

　(iii) describe how certain stem cells in the bone marrow develop into erythrocytes. (3)

(c) State *two* potential benefits of using therapeutic cloning to produce tissues and organs for transplantation. (2)

8 (a) Explain why people infected with HIV may develop AIDS. (3)

(b) Explain why it is difficult to control the spread of HIV, particularly in less economically developed countries.

In this question one mark is available for the quality of spelling, punctuation and grammar. (7)

Quality of written communication (QWC: 1)

9 (a) Severe acute respiratory syndrome (SARS) is caused by a virus.

In 2003, health officials feared that this virus could have given rise to a new, highly infectious disease that might have led to a pandemic.

　(i) State what is meant by the term *pandemic*. (1)

　(ii) Suggest why the **spread** of a newly emerged infectious disease is likely to be difficult to control. (2)

(b) Complete the following description of cancer by using appropriate words from the list to fill in the gaps.

meiosis　mitosis　tumour　malignant

mutation　proto-oncogenes　translation

benign　transcription　regulators　cyst

Cancer is a disease that starts in our cells. A .... occurs in special genes called .... that are found in all normal cells.

This leads to a lack of control of ....

The uncontrolled growth of cells forms a lump called a ....

Some lumps are .... and may not cause any problems. Other lumps are .... and contain cells that can spread into other body tissues causing severe damage.

(6)

10 MISTLETOE IN MEDICINE

Mistletoe, *Viscum album*, is a parasitic plant that grows on other plants such as apple trees. The mystical properties of mistletoe have been recognised since the time of the druids, who called it 'all heal'.

As an ancient herbal remedy, many people have researched and documented the medical uses of mistletoe and its extracts since the 18th century. Mistletoe tea, dried leaves soaked in water, is available in many continental pharmacies where it is used to alleviate a range of conditions including hypertension.

In the 1920s, Rudolph Steiner made an extraordinary prediction about the plant. He believed that, as a parasitic plant, mistletoe should have medicinal value as an anti-cancer agent and he formulated a variety of extracts for anti-cancer use. The prediction is based on the assumption of homeopathy that 'like cures like'. He perceived cancers to be parasitic on the human body and hence extracts of a parasitic plant, such as mistletoe, should be suitable as homeopathic remedies for cancer.

Such extracts are still being manufactured, for example Iscador. They are widely used as part of complementary and alternative medicine (CAM) in cancer treatment. In Germany, for example, more than £20 million is spent each year on mistletoe extracts to fight cancer

Several different chemicals have been isolated and purified from such mistletoe extracts. Some have been shown to have potential anti-cancer properties. One such group of chemicals are the mistletoe lectins. Mistletoe lectin 1 has been shown to inhibit protein synthesis at the ribosome level.

It has also been shown to be a possible modifier of the immune response. It is thought that many cells in all of us become potentially cancerous. The immune system normally identifies and destroys these cells before they develop into tumours. Mistletoe lectin 1 is thought to enhance this process by activating macrophages and stimulating the multiplication of T helper lymphocytes.

Trials on mistletoe extracts have reported a reduction in the mass of solid tumours and an improved quality of life. As with many other CAM therapies, these trials are often unreliable. In a review of 11 trials, only one was found to be reliable and this was the only one reporting no improvement from using mistletoe.

The continued wide use of CAM therapies in cancer treatment is emphasised by a July 2000 study in the *Journal of Clinical Oncology*. This reported that, of 453 cancer patients, 69% had used at least one CAM therapy.

(a) How does a cell from mistletoe differ from a human cell such as a white blood cell? (3)

(b) Plant extracts have been widely used in the treatment of cancers. Some of these are used in complementary therapies. In the study of July 2000, how many people had used at least one CAM therapy? (2)

Lectins from mistletoe extract were shown to inhibit protein synthesis at the ribosome.

(c) Suggest why the inhibition of protein synthesis may lead to cell death and a reduction in the mass of solid tumours. (1)

One possible role of lectins may be to stimulate macrophages and lymphocytes.

(d) Explain how cancer develops, and describe the role of lymphocytes in preventing the development of cancer. (7)

UNIT 4

Respiration

✓ Quick check 1

✓ Quick check 2

✓ Quick check 3

✓ Quick check 4

ATP

ATP is a molecule with high energy potential. It is easily transported in cells, easily made by addition of phosphate to ADP and easily **hydrolysed** to release the potential energy. Hydrolysis of the final phosphate is the main energy source: 30.5 kJ of energy are released when this bond is hydrolysed. Some energy will be lost as heat, but the majority is used by cells to do work.

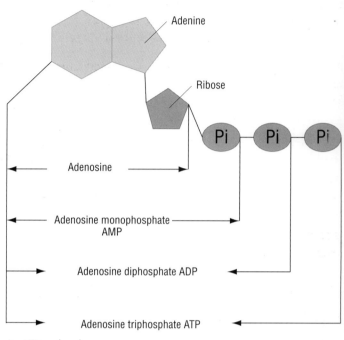

An ATP molecule

Use of ATP

ATP is recycled in cells – it is replaced as it is used.
The energy from hydrolysis of ATP is released in small amounts allowing it to be used in metabolic processes such as active transport and muscle contraction.

ATP synthesis may occur in the cell cytoplasm or within the inner mitochondrial membranes, or in the thylakoid membranes of chloroplasts.

Glycolysis

Glycolysis occurs in the cell cytoplasm, and is the process where glucose (6C) is split to produce two molecules of pyruvate (3C), reduced NAD *and* also ATP by substrate level phosphorylation.

Link reaction

Pyruvate is a high energy molecule. If oxygen is present pyruvate can be broken down further by entering the mitochondria. It is **dehydrogenated** by enzymes and NAD picks up the hydrogen, becoming reduced NAD. Decarboxylase enzymes break it down to release CO_2 and a 2C molecule, acetyl coenzyme A, is formed.

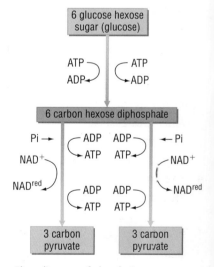

Flow diagram of glycolysis

The link reaction

Krebs cycle

The **Krebs cycle** is a series of enzyme-controlled steps in the mitochondrial matrix. You do not need to know all of the steps. The hydrogen released reduces the co-enzymes NAD and FAD.

Acetyl coenzyme A combines with 4C oxaloacetate and forms an intermediate 6C molecule, citrate. This is quickly **decarboxylated** (releasing CO_2) and dehydrogenated by a series of enzyme-linked steps to regenerate the 4C molecule and complete the cycle. The hydrogen released reduces NAD and FAD. The cycle can now start again. One ATP is formed directly for each cycle by substrate level phosphorylation.

Oxidative phosphorylation

ATP is produced using the potential energy in reduced NAD and FAD. The process is called **oxidative phosphorylation**. It occurs on the cristae of the mitochondria where all the necessary enzymes and carriers are found.

- Reduced NAD and FAD lose their hydrogen atoms.
- Electrons from the hydrogen are transferred along a series of **cytochromes** (electron carriers) embedded in the membrane of the cristae. Each cytochrome is reduced as it accepts the electron and is oxidised as it passes it on (redox). The energy released is captured as ATP – three molecules from reduced NAD and two from reduced FAD.
- The final acceptor for electrons is oxygen which also picks up hydrogen ions to form water.

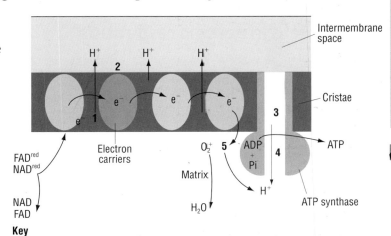

The Krebs cycle

Key
1 Oxidation and reduction of NAD and FAD
2 Electrons pass along the carriers, and protons (H^+) are pumped into inter-membrane space
3 Protons diffuse down proton gradient back into matrix
4 ATP synthase phosphorylates ADP into ATP
5 Electrons move along membrane carriers to final carrier molecule oxygen

Diagram of oxidative phosphorylation

QUICK CHECK QUESTIONS

1 Why is ATP referred to as energy currency?
2 Describe and explain the difference between the different types of phosphorylation.
3 Why is coenzyme A needed in the link reaction?

4 Name the products of Krebs cycle regardless of the original source of acetyl coenzyme A.
5 Explain the meaning of a redox reaction and how it can be used to describe the passage of electrons in oxidative phosphorylation.

UNIT 4

Chemiosmosis

Key words

- chemiosmosis
- electron transport chain
- respiratory substrate
- respiratory quotient
- proton (H+) gradient
- respirometer

✓ *Quick check 1*

✓ *Quick check 2*

✓ *Quick check 3*

ATP is produced by the process of **chemiosmosis**. The enzyme ATP synthase is required and a hydrogen ion concentration gradient is needed.

The process

- The energy from transferring electrons in the carrier chain pumps protons (H⁺) into the intermembrane space of the mitochondria.
- An electrochemical gradient (proton gradient) is produced and the H⁺ diffuse back to the matrix through protein channels in the ATP synthase.
- Proton movement through the channels fuels the conversion of ADP to ATP by rotating the enzyme ATP synthase.
- The H⁺ and electrons now recombine and join with oxygen to form water.

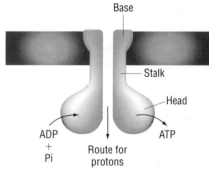

Stalk particle

Experimental evidence for chemiosmosis

Peter Mitchell established that build up of a hydrogen ion gradient was necessary for ATP production, and that energy released from the transfer of electrons along the chain was used to pump H⁺ into the intermembrane space.

Other researchers showed that ATP formation only occurred if there was the enzyme ATP synthase *and* an inter-membrane space.

ATP yield in aerobic respiration

In aerobic respiration, ATP is produced at various steps but most comes from oxidative phosphorylation in the electron transport chain. A steady supply of oxygen is needed as the final acceptor. A theoretical total of 32 ATP molecules are formed:

- glycolysis – two by substrate phosphorylation
- Krebs cycle – one per turn of the cycle giving a total of two
- oxidative phosphorylation – 28.

Another six may be formed if the two reduced NAD from glycolysis move into the mitochondria. However, only rarely are all possible ATP molecules formed as all conditions must be at their optimum. Also, some ATP may be used up, e.g. to shunt the reduced NAD from glycolysis.

ATP yield in anaerobic respiration

In the absence of oxygen as the final electron acceptor, the breakdown of glucose cannot proceed beyond glycolysis. However, the reduced NAD must be restored to NAD or glycolysis will also stop and no further ATP is formed.

Anaerobic respiration allows glycolysis to continue by recycling reduced NAD molecules. The hydrogen from reduced NAD is passed on to pyruvate, reducing it to lactate, and the NAD can be reused. This results in a net gain of two ATP molecules.

Respiratory substrate

Substrates other than glucose can be broken down and used to produce ATP. Lipids are broken down into fatty acids and glycerol. The fatty acids are converted into many 2C acetyl molecules which combine with coenzyme A and enter Krebs. Glycerol can also be used. The result is a greater number of ATP molecules per gram from fat than from glucose since lipids contain more hydrogen atoms than carbohydrates.

Proteins are broken down to amino acids and, in the liver, amino groups are removed from excess amino acids by deamination. The resulting organic molecule then enters Krebs. The amine group enters the ornithine cycle and is converted to urea. The organic molecule produces a number of ATP molecules.

✔*Quick check 4*

Respiratory quotient (RQ)

When ATP is formed from a respiratory substrate O_2 is used and CO_2 is produced. The ratio between these can be used to determine the exact substrate being used by cells and whether it is being broken down in aerobic or anaerobic respiration.

$$RQ = \frac{\text{volume of } CO_2 \text{ produced}}{\text{volume of } O_2 \text{ consumed}}$$

Substrate	Energy value (kJ g^{-1})	RQ
Glucose	16	1
Lipid	39	0.7
Protein	17	0.9

With a mixture of lipids and glucose respired an RQ of 0.85 occurs. An RQ value of greater than 1 means some anaerobic respiration is happening.

Measuring RQ

A **respirometer** is a simple method of measuring gas taken in and gas produced from a living organism, such as respiring seedlings or small invertebrates. This allows the RQ to be calculated.

A respirometer can also be used to investigate the rise in respiration rate with increasing temperatures. The rate of production of carbon dioxide by yeast cells can be used to investigate the effect of using sugars, such as maltose or lactose, as substrates.

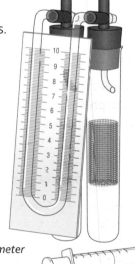

A respirometer

✔*Quick check 5*

✔*Quick check 6*

Key definition

Respiratory quotient or RQ is the ratio between oxygen used up and carbon dioxide given out. It can be used to work out what respiratory substrate is being respired. It will also indicate whether respiration is aerobic, when the value is 1 or less than 1, or anaerobic, when the value is greater than 1 or infinity.

QUICK CHECK QUESTIONS

1 Explain how mitochondrial structure is related to ATP production.

2 Describe the process of ATP production by chemiosmosis.

3 Why is it important to recycle NAD in anaerobic respiration?

4 What is a respiratory substrate?

5 Give reasons why the RQ value calculated in the table above may not be exactly 1 or 0.7 or 0.9.

6 Explain how a respirometer may be used to calculate RQ.

Exercise

Key words

- aerobic
- steroids
- anaerobic
- degenerative
- carbohydrate loading
- recombinant

✓*Quick check 1*

Aerobic exercise is important for overall body health.
Recommended exercise levels are three 30-minute sessions weekly to reduce obesity and the risk of cardiovascular disease. However, a regular programme of exercise, up to 200 minutes per week, is desirable.

Short-term consequences of exercise

Effect	Cause	Consequence
Increased heart rate Increased stroke volume	Increased adrenaline and sympathetic nervous stimulation	Increased O_2 for aerobic respiration and increased removal of CO_2
Increased cardiac output	Muscle cells secrete nitric oxide as O_2 falls causing vasodilation of muscle arterioles and increased flow back to the heart This increases blood pumped out at systole	
Blood redistribution away from gut and towards muscles and heart	Vasodilation of arterioles to muscles and vasoconstriction to non-essential organs	O_2 delivered to essential organs only
Dilation of arterioles in skin	Adrenaline and vasodilation	Allows excess heat to be lost and cool the body during exercise
Increased breathing rate and tidal volume	Increased CO_2 from respiring cells lowers blood pH Chemoreceptors in medulla and aortic arch are stimulated and rate increases	Increases O_2 uptake for aerobic respiration

Long-term consequences of regular exercise (training)

Effect	Cause	Benefit/consequence
Increased vascularisation of muscles and lungs	Faster diffusion rates with more capillaries	Increased uptake and delivery of O_2 to mitochondria in cells
Increase in size of muscle fibres, more slow-twitch fibres	Sustained muscle contraction	Increase in sustained aerobic exercise
More mitochondria	More ATP produced	
Increased glycogen store	Increased substrate	
Increased myoglobin	Increased O_2 store	
Higher VO_2max	Sustained aerobic respiration	Increase in breathing rate and delivery of O_2 to cells Decrease in recovery time
Increased heart muscle leading to increased stroke volume	Increased O_2 delivery, and fat stores can be utilised since they are only broken down aerobically	Increased substrate for aerobic respiration
Decreased resting heart rate and blood pressure	Reduced heart muscle contraction	Reduces strain on heart
Increased maximum breathing rate and vital capacity	Increase in amount and rate of O_2 uptake	Increase in duration of aerobic respiration

Enhancing performance

- **Carbohydrate loading**. Glycogen stores in muscle and liver are deliberately increased. Athletes achieve this by first depleting glycogen stores for a few days before an event. They will then consume large amounts of carbohydrates just before the event.

- **Recombinant DNA** contains genes from different sources combined with the original DNA. It is usually manipulated by genetic engineering. Recombinant erythropoietin is formed in this way and provides a source of erythropoietin which is free of disease since t is not from animal sources.

- Blood doping, including blood transfusions and the use of **recombinant human erythropoietin (RhEPO)** to increase O_2 transport. Both methods are used to increase red blood cell counts and so improve performance. Blood is removed and stored months before an event. The body will replace the lost blood. Just before the event the stored blood is warmed and replaced, increasing the blood volume. RhEPO is injected. It stimulates the bone marrow to produce more red blood cells and so the athlete benefits from increased red cells. ✓*Quick check 4*

- **Steroid enhancement**. Artificial anabolic steroids are used to increase protein synthesis and increase muscle bulk before an event. This provides an unfair advantage and makes the athlete more aggressive and competitive. ✓*Quick check 5*

- **Legally enhancing**. Extra creatine phosphate can be taken by athletes. The phosphate will readily combine with ADP to form ATP, providing an energy source. The creatine will be reformed into creatine phosphate when there is a lowered energy demand in the muscles. This is a system naturally occuring in muscle cells that is simply being manipulated.

Key definition

Recombinant DNA contains genes from different sources combined with the original DNA. It is usually manipulated by genetic engineering. Recombinant erythropoietin is formed in this way and provides a source of erythropoietin which is free of disease since it is not from animal sources.

Examiner tip

A drop in blood volume or O_2 levels causes kidney glomerular cells to produce the hormone erythropoietin which stimulates red blood cell production.

Examiner tip

Only carbohydrate loading is legal in competitive sport. There are medical problems associated with enhancement using other methods.

QUICK CHECK QUESTIONS

1 Why is an exercise program designed to raise your heart rate to 70% of your maximum?

2 Explain the effect of nitric oxide secretion.

3 Summarise the main benefits of training in athletes.

4 What health problems do you think may be associated with illegal enhancement?

5 Describe how RhEPO gives an athlete an unfair advantage.

Making haemoglobin

Haemoglobin synthesis

Haemoglobin is the respiratory pigment in red blood cells. It is a globular protein with a quaternary structure. The four haem groups each combine reversibly with O_2 molecules forming oxyhaemoglobin.

There are two genes coding for haemoglobin protein (these are on chromosomes 11 and 16 in the nucleus). The triplet sequence of DNA bases in a gene forms the template for mRNA synthesis. Each triplet of bases codes for one amino acid. The code is degenerative since some amino acids are coded for by more than one triplet. Each triplet is distinct and non-overlapping. There are three stop triplets signalling the end of the gene. The triplet for the amino acid methionine is the start triplet.

Key definition

DNA is made up of many sequences of bases forming a triplet code. This sequence consists of both coding DNA which codes for proteins and non-coding DNA which does not code for proteins. When a length of DNA is transcribed to form mRNA, both coding and non-coding sequences are copied. Once the mRNA reaches the ribosomes, enzymes edit the strand and cut out the non-coding DNA called introns leaving the mRNA shorter. It now consists only of the coding DNA sequences called exons.

Protein synthesis occurs on **ribosomes**. Initially the alpha and beta polypeptide chains are formed. Haemoglobin is then formed by combining the polypeptides (2 alpha and 2 beta) with haem prosthetic groups. The final haemoglobin protein is a three-dimensional quaternary protein.

Transcription

This occurs in the nucleus. The mRNA template is copied from the gene.
Once formed the mRNA moves into the cytoplasm and attaches to ribosomes.

Translation

The sequence of **codons** (mRNA triplets) is interpreted into an amino acid sequence using ribosomes and tRNA. The tRNA attaches to its correct amino acid in the cytoplasm and brings it to the correct mRNA codon. The ribosome holds the amino acids in place while a peptide bond forms between adjacent amino acids and builds the polypeptide (see diagram on next page).

✓Quick check 1

Gene mutation

A **gene mutation** is a permanent change to the DNA sequence. The resulting protein can have a different tertiary structure and different properties.

Sickle cell anaemia is a mutation in one single base in the β haemoglobin gene. This is called a point mutation. The resulting triplet codes for the amino acid valine instead of for glutamate. This changes the structure and properties of haemoglobin. It becomes insoluble at low pO_2 causing the red blood cells to become sickle shaped and leads to shortage of oxygen: hypoxia.

✓Quick check 3

1.

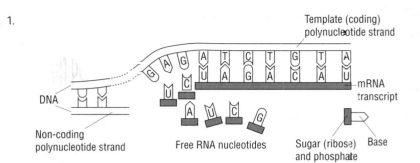

DNA

Non-coding
polynucleotide strand

Free RNA nucleotides

Template (coding)
polynucleotide strand

mRNA
transcript

Sugar (ribose)   Base
and phosphate

4.

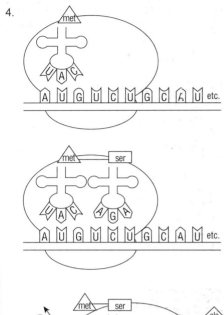

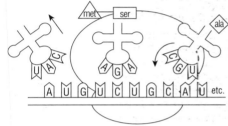

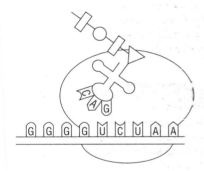

2.

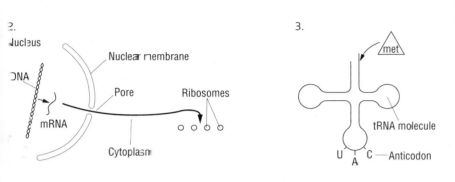

Nucleus

Nuclear membrane

DNA

Pore

Ribosomes

mRNA

Cytoplasm

3.

met

tRNA molecule

U   C — Anticodon
   A

Key

1   Transcription of mRNA from DNA
2   mRNA moving out of nucleus to the ribosomes
3   tRNA being activated with an amino acid (met) from the cytoplasm
–   Sequence of diagrams showing translation

DNA repair

DNA replication frequently results in mistakes that would lead to severe
problems or death. However, DNA polymerase checks the DNA copy.
Other enzymes cut out the mistake if found and DNA polymerases and
ligases repair it.

✔ *Quick check 4*

Cellular ageing

Telomeres at the ends of chromosomes protect the genes and regulate
cell divisions.
Over time, telomeres shorten leaving the genes exposed to damage.
Once the DNA is not repairable, cell death (apoptosis) is triggered.
Stem cells maintain their telomeres using telomerase and so can
continue to divide.

✔ *Quick check 5 and 6*

QUICK CHECK QUESTIONS

1   Draw a flow chart to show the steps involved in protein
    synthesis.

2   Describe all the roles carried out by RNA molecules.

3   Explain the term hypoxia and why ATP production will
    fall.

4   Describe the similarities and differences between DNA
    repair and recombinant DNA.

5   What role may telomerase play in cancer cells?

6   Explain what apoptosis is and how it can benefit the
    cells.

Haemoglobin and myoglobin

Key words

- myoglobin
- oxyhaemoglobin
- oxyhaemoglobin dissociation curve
- Bohr shift
- oxygen debt
- EPOC

Hint

Haemoglobin is a complex three-dimensional protein made up of four polypeptide chains each with a haem group attached. It combines with four O_2 molecules when the partial pressure of oxygen (pO_2) is high, to form oxyhaemoglobin.

✓ *Quick check 1 and 2*

Examiner tip

Remember in your alveoli O_2 levels are high and O_2 readily diffuses and loads. The haemoglobin in cells leaving the lungs will be carrying its maximum number of oxygen molecules. In tissues such as active muscles, O_2 levels are low and O_2 is readily released. This is reflected in the sigmoid shape of the oxyhaemoglobin dissociation curve.

✓ *Quick check 3*

Myoglobin

Myoglobin is a protein which acts as an O_2 store in muscle cells. It has only one polypeptide chain and haem group. It forms oxymyoglobin when it picks up O_2 from **oxyhaemoglobin**. It releases O_2 when levels fall, e.g. in the muscles after exercise. It has a higher affinity for oxygen than haemoglobin and so will pick up oxygen when oxyhaemoglobin is releasing it.

Key definition

Partial pressure of oxygen is the pressure the oxygen exerts in a gas mixture. It is given as a measure of the amount of oxygen present. It is high in the lungs since air entering has a high level of oxygen. In the tissues the partial pressure of oxygen is low since the respiring tissues have used large amounts of oxygen.

Oxyhaemoglobin dissociation curve

This curve shows the relationship between the partial pressure of oxygen and the percentage saturation of haemoglobin.

At low pO_2 haemoglobin loads less easily and offloads more easily producing the shallow part of the S curve. The haem group does not pick up oxygen readily because it is in the centre of the molecule. However, once the first oxygen has loaded, the shape of the haemoglobin molecule changes, allowing more oxygen molecules to bind more readily.

The result is that at high pO_2 there is high affinity for O_2 and it is difficult to offload O_2, producing the flat part of the S curve. The steep part of the curve corresponds to pO_2 values in respiring tissues.

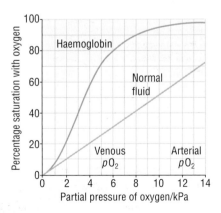

Oxyhaemoglobin dissociation curve

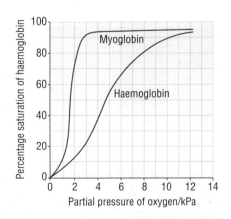

Oxygen dissociation curve for haemoglobin and myoglobin

Bohr shift

In respiring tissues CO_2 levels are high and this forms carbonic acid readily in solution. Carbonic acid moves the O_2 dissociation curve to the *right*, because as it dissociates it releases H^+. Haemoglobin acts as a buffer to the H^+ and forms haemoglobinic acid (HHb). This change displaces the oxygen from the oxyhaemoglobin. We say haemoglobin's affinity for O_2 is lowered. So O_2 is more easily released if respiration rates are high.

✓ *Quick check 4*

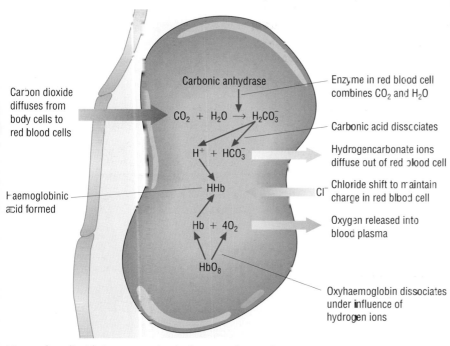

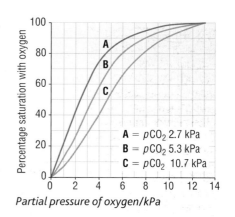

How carbon dioxide is converted to hydrogencarbonate ions

Oxygen debt

During strenuous exercise insufficient oxygen is available so ATP is produced anaerobically. The resulting lactic acid must be broken down in the liver after the exercise is over. The extra O_2 required to do this above normal body needs (**EPOC**) is called the **oxygen debt**. During this time the body's store of oxyhaemoglobin, oxymyoglobin, creatine phosphate and ATP are also restored.

✓ *Quick check 5*

✓ *Quick check 6*

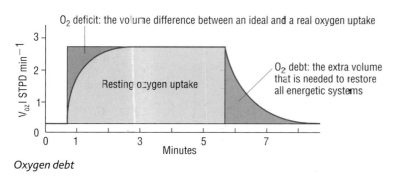

Oxygen debt

QUICK CHECK QUESTIONS

1 State the differences between myoglobin and haemoglobin.
2 What does oxyhaemoglobin signify?
3 Describe how the oxyhaemoglobin dissociation curve helps explain transport and delivery of oxygen to respiring cells.

4 How does an increase in carbon dioxide lower the affinity haemoglobin has for oxygen?
5 What are the roles of myoglobin and creatine phosphate?
6 Explain how you would calculate the EPOC value using a graph such as the one above.

Skeletal muscle

Key words

- myogenic
- sarcoplasm
- sarcolemma
- actin
- myosin

Muscle

There are three types of muscle tissue: cardiac, skeletal and smooth muscle. They all contain the proteins actin and myosin, which form contractile fibrils in the cell cytoplasm. All muscle can contract and relax using ATP.

Cardiac muscle is a special kind found only in the heart. It never tires, is **myogenic** and has the ability to beat constantly.

Smooth muscle (involuntary muscle) is found in many internal organs and structures such as the artery wall and the gut. It contracts when stimulated by the autonomic nervous system.

✓ *Quick check 1*

Skeletal muscles are attached to the skeleton and cause movement. Skeletal muscle is voluntary muscle. It responds to nervous stimulation by the somatic nervous system.

Skeletal muscle structure

Each muscle cell contains **sarcoplasm**, and has many mitochondria. The cell is surrounded by a **sarcolemma** (membrane).

This membrane has infolds called T tubules which link to the sarcoplasmic reticulum and form a membrane network through the sarcoplasm. The T tubules contain Ca^{++} and the sarcoplasm contains **actin** and **myosin** myofibrils which are all vital to muscle contraction.

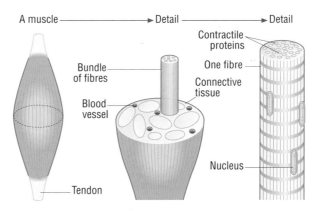

✓ *Quick check 1 and 2*

There are 3 types of muscle, each with a different function

Sarcomere

The myofibrils contain actin and myosin arranged in a regular pattern making up the sarcomere. The thin filaments are actin and the thick filaments between are myosin. Actin is made of three proteins twisted together, actin, tropomyosin and troponin. Myosin is bundles of proteins with a golfclub-shaped head used to bind to the actin and to ATP. The myosin head is an ATPase which can hydrolyse ATP releasing energy.

Key definition

A sarcomere is the length or unit of the muscle cell. It corresponds to the internal bands formed by the actin and myosin. There are two membranes on which the actin attaches and an inner membrane on which the myosin attaches.

The prefix sarco- refers to the muscle, so sarcoplasm is the cytoplasm within the muscle cell.

✓ *Quick check 2*

Sliding filament theory

Contraction of a muscle cell causes the sarcomere to shorten. Actin slides between the myosin reducing the overall length. This temporarily reduces the width of the H band until the muscle relaxes.

Contraction of skeletal muscle

- Nervous stimulation is received by the muscle cell.
- The impulse travels down the T tubules to the sarcoplasmic reticulum.
- Ca^{2+} ions are released into the cytoplasm from the sarcoplasmic reticulum.
- The calcium ions bind to the troponin and remove the tropomyosin that is acting as a blocking molecule in the actin binding sites.
- The myosin heads can now attach to the exposed actin binding sites.
- The Ca^{2+} ions stimulate an enzyme which frees the myosin head from ADP so that the head is able to attach.
- The myosin head changes position and 'rows' the actin so the filaments slide past each other. This is called the power stroke.
- Attachments break and reattach further along so actin is pulled along.
- ATP is now needed to hydrolyse and release the energy needed to return the myosin head back into its original position. It will attach to the remaining ADP molecule.
- Ca^{2+} ions leave the binding site and contraction can begin again.
- While there is nervous stimulation Ca^{2+} will be released and the cycle will be repeated. While ATP and Ca^{2+} are present the sarcomere will continue to contract.

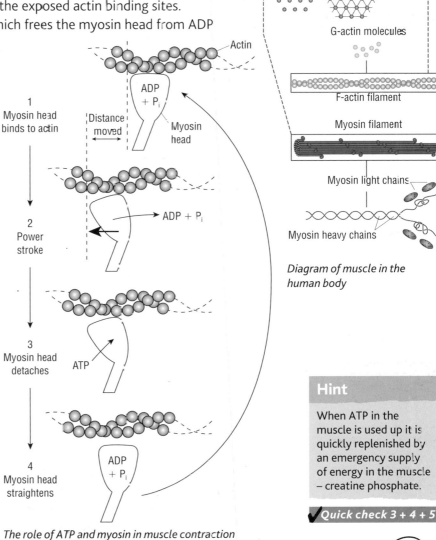

The role of ATP and myosin in muscle contraction

Diagram of muscle in the human body

Examiner tip

You should be able to describe the sliding filament theory of muscle contraction and identify all the key structures and terms in a muscle cell.

Hint

When ATP in the muscle is used up it is quickly replenished by an emergency supply of energy in the muscle – creatine phosphate.

✔ *Quick check 3 + 4 + 5*

QUICK CHECK QUESTIONS

1 What is the difference between a muscle fibre and a myofibril?
2 What is a sarcomere? Annotate a diagram to help you explain.
3 Describe the role of ATP in muscle contraction.
4 What is the role of Ca^{2+} in the process of muscle contraction?
5 Explain the importance of the power stroke.
6 Creatine phosphate can be used to regenerate ATP from ADP. What happens when the muscles' supply of creatine phosphate is used up?

Human reproductive systems

Key words

- ovaries
- follicle
- testes
- seminiferous
- mammary glands
- fallopian tube
- sperm duct
- prostate gland

The female reproductive organs have a number of roles:

- producing female gametes (oocytes) in the ovaries
- the site of fertilisation (Fallopian tube) and fetal development (uterus)
- producing female sex hormones – oestrogen and progesterone.

✔ *Quick check 1*

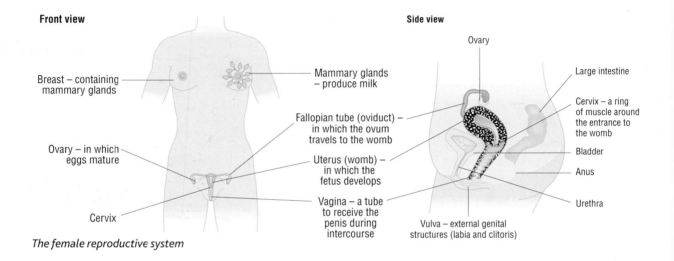

Front view

Breast – containing mammary glands

Ovary – in which eggs mature

Cervix

Mammary glands – produce milk

Fallopian tube (oviduct) – in which the ovum travels to the womb

Uterus (womb) – in which the fetus develops

Vagina – a tube to receive the penis during intercourse

Side view

Ovary

Large intestine

Cervix – a ring of muscle around the entrance to the womb

Bladder

Anus

Urethra

Vulva – external genital structures (labia and clitoris)

The female reproductive system

The male reproductive organs also have several roles:

- producing male gametes – sperm
- producing male sex hormone – testosterone
- introducing sperm into the female for fertilisation.

✔ *Quick check 2*

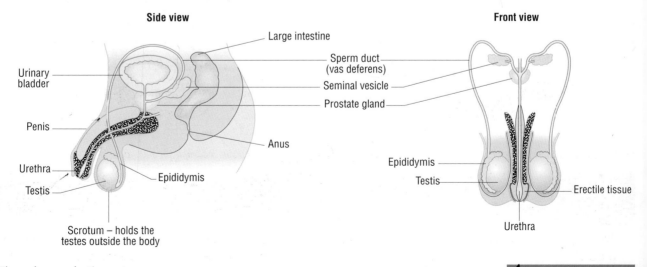

Side view

Urinary bladder

Penis

Urethra

Testis

Epididymis

Scrotum – holds the testes outside the body

Large intestine

Sperm duct (vas deferens)

Seminal vesicle

Prostate gland

Anus

Front view

Epididymis

Testis

Erectile tissue

Urethra

The male reproductive system

✔ *Quick check 3*

Ovary and testis

The **ovaries** are a matrix containing many primary **follicles**. The follicles in an ovary are at different stages of development and maturity.

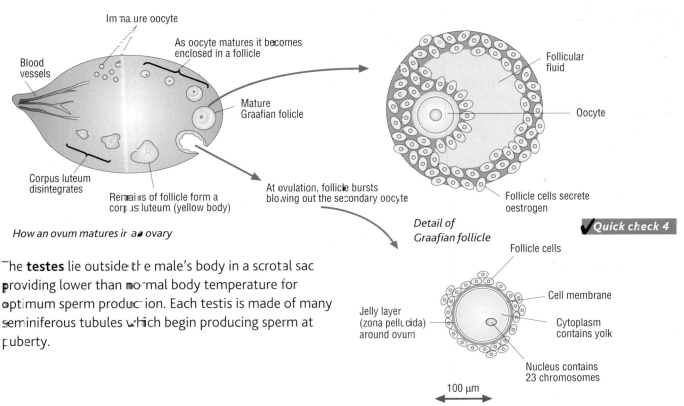

Immature oocyte

Blood vessels

As oocyte matures it becomes enclosed in a follicle

Mature Graafian folicle

Corpus luteum disintegrates

Remains of follicle form a corpus luteum (yellow body)

At ovulation, follicle bursts blowing out the secondary oocyte

How an ovum matures in an ovary

Follicular fluid

Oocyte

Follicle cells secrete oestrogen

Detail of Graafian folicle

✓ **Quick check 4**

Follicle cells

Cell membrane

Jelly layer (zona pellucida) around ovum

Cytoplasm contains yolk

Nucleus contains 23 chromosomes

100 μm

Detail of secondary oocyte

The **testes** lie outside the male's body in a scrotal sac providing lower than normal body temperature for optimum sperm production. Each testis is made of many seminiferous tubules which begin producing sperm at puberty.

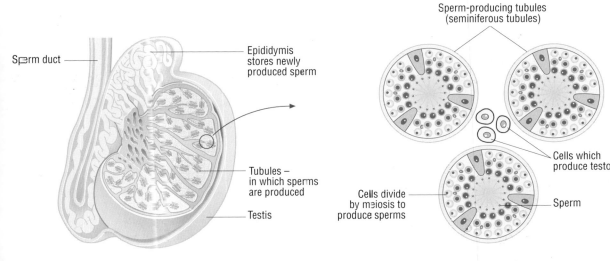

Sperm duct

Epididymis stores newly produced sperm

Tubules – in which sperms are produced

Testis

Section through a testis

Sperm-producing tubules (seminiferous tubules)

Cells which produce testo

Cells divide by meiosis to produce sperms

Sperm

Section through the seminiferous tubules

QUICK CHECK QUESTIONS

1 Why are the sex organs also described as endocrine organs?
2 Describe the roles of the male and female reproductive organs.

3 Explain how the differences in the roles of the female and male reproductive organs are reflected in the differences in their structures.
4 What is the difference between the follicle and the egg cell?

Hormones, gametogenesis and fertilisation

Key words

- gametogenesis
- oocyte
- acrosome
- meiotic
- zygote
- spermatocyte
- FSH
- LH
- gonadotrophin

Gametogenesis

Production of the female gamete (**oogenesis**) begins in the fetus. The diploid germ cells divide to form oogonia. Meiotic cell division begins but stops at prophase 1 resulting in primary oocytes. These are surrounded by ovarian cells and form a follicle. By puberty the ovaries contain around 200 000 primary oocytes.

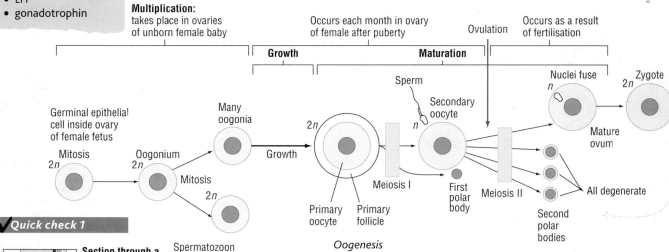

Multiplication: takes place in ovaries of unborn female baby

Occurs each month in ovary of female after puberty

Growth   **Maturation**

Ovulation

Occurs as a result of fertilisation

Germinal epithelial cell inside ovary of female fetus

Mitosis $2n$

Oogonium $2n$

Mitosis

$2n$

Many oogonia

Growth

Primary oocyte   Primary follicle $2n$

Meiosis I

First polar body

Sperm n

Secondary oocyte

Meiosis II

Second polar bodies

All degenerate

Nuclei fuse n

Mature ovum

Zygote $2n$

Oogenesis

✓ *Quick check 1*

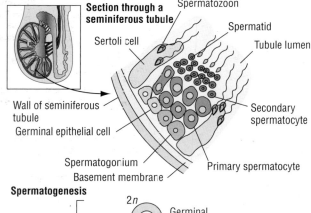

Section through a seminiferous tubule

Spermatozoon

Sertoli cell

Spermatid

Tubule lumen

Wall of seminiferous tubule

Germinal epithelial cell

Secondary spermatocyte

Spermatogonium

Basement membrane

Primary spermatocyte

From puberty, each month oocytes progress to telophase 1, although only one continues to mature to become a secondary oocyte. This secondary oocyte continues meiosis to metaphase 2 and is released at ovulation. The secondary oocyte is a large haploid cell surrounded by the zona pellucida and corona radiata.

Production of sperm called **spermatogenesis** occurs in the **seminiferous** tubules. These are lined inside with diploid spermatogonia. These divide by mitosis and grow into primary spermatocytes. The first meiotic division results in secondary spermatocytes. The second meiotic division produces haploid spermatids that grow and differentiate into sperm cells supported by nurse cells called sertoli cells.

The sperm cells consist of:

- a head with haploid nucleus and an **acrosome** tip of hydrolytic enzymes
- a midpiece containing numerous mitochondria
- a tail consisting of contractile filaments.

✓ *Quick check 2*

Role of hormones

Gametogenesis is controlled by hormones from the hypothalamus and anterior lobe of the pituitary gland. In turn, these hormones stimulate the ovaries or testes to produce hormones.

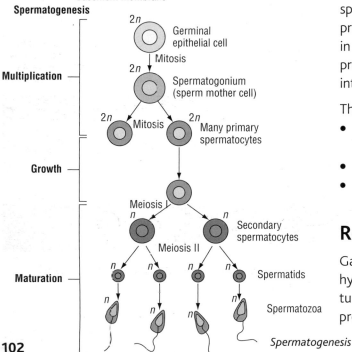

Spermatogenesis

Multiplication

$2n$ Germinal epithelial cell

Mitosis

$2n$ Spermatogonium (sperm mother cell)

$2n$ Mitosis $2n$ Many primary spermatocytes

Growth

Meiosis I

n n Secondary spermatocytes

Meiosis II

n n n n Spermatids

Maturation

n n n n Spermatozoa

Spermatogenesis

Hormones in menstrual cycle

Hormone	Producing organ	Effect
Gonadotrophin	Hypothalamus	Stimulates anterior pituitary to produce two hormones
FSH (follicle-stimulating hormone)	Anterior pituitary	Binds to follicle cells and stimulates follicle to mature and produce oestrogen
LH (luteinising hormone)		Stimulates ovulation and development of corpus luteum
Oestrogen	Mature follicle cells	Rising levels inhibit FSH and initially LH followed by LH surge. Causes endometrium to begin thickening
Progesterone	Corpus luteum	Rising levels inhibit FSH and stimulate secretory phase of endometrium development

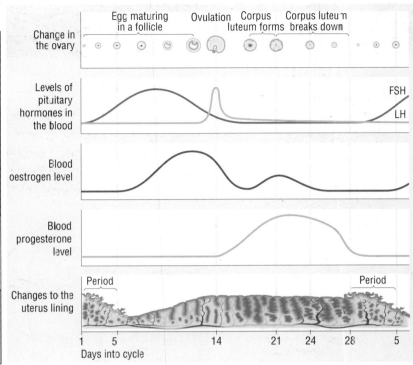

Changes during the menstrual cycle

✓ *Quick check 3*

Fertilisation

Fertilisation may result if a sperm is released into the female reproductive tract to meet a secondary oocyte in the Fallopian tube.

- The sperm reaching the oocyte, binds to a receptor and releases acrosome enzymes to allow it to penetrate the zona pellucida.
- At the cell membrane it binds to another receptor to stimulate changes in the zona pellucida and stop entry of other sperm.
- The haploid male nucleus stimulates the oocyte to complete meiosis and divide (the second cell produced becomes a polar body).
- The male nucleus enters and fuses with the remaining oocyte nucleus.
- A diploid **zygote** is formed. The diploid chromosome number is restored. Since chromosomes of each pair are from two different parents, there will be new combinations of alleles in the offspring.

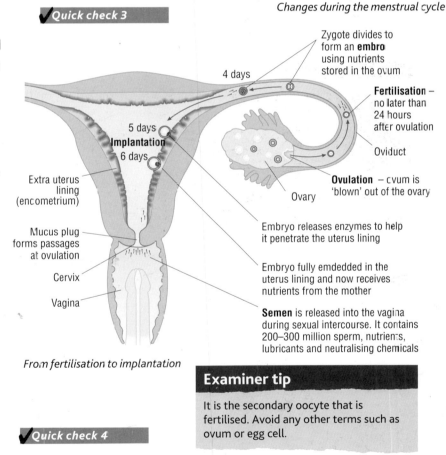

From fertilisation to implantation

Zygote divides to form an **embryo** using nutrients stored in the ovum

4 days

Fertilisation – no later than 24 hours after ovulation

Oviduct

Ovulation – ovum is 'blown' out of the ovary

Ovary

5 days
Implantation
6 days

Extra uterus lining (endometrium)

Mucus plug forms passages at ovulation

Cervix

Vagina

Embryo releases enzymes to help it penetrate the uterus lining

Embryo fully embedded in the uterus lining and now receives nutrients from the mother

Semen is released into the vagina during sexual intercourse. It contains 200–300 million sperm, nutrients, lubricants and neutralising chemicals

Examiner tip

It is the secondary oocyte that is fertilised. Avoid any other terms such as ovum or egg cell.

✓ *Quick check 4*

QUICK CHECK QUESTIONS

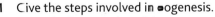

1 Give the steps involved in oogenesis.
2 State the importance of the acrosome.
3 Draw a flowchart to show the interrelationship of hormones in the menstrual cycle.

4 Describe the steps involved and the significance of fertilisation.

Pregnancy and contraception

Key words

- implantation
- trophoblast
- chorion
- human chorionic gonadotrophin (HCG)
- follicle-stimulating hormone (FSH)
- oestrogen
- progesterone
- oxytocin
- prolactin
- birth-control pill
- intrauterine device (IUD)

✓*Quick check 1*

Examiner tip

Human chorionic gonadotrophic hormone (HCG) is secreted to prevent the corpus luteum from breaking down and allows progesterone production to continue. Progesterone maintains the uterus lining and so maintains the pregnancy and inhibits **follicle-stimulating hormone (FSH)**.

✓*Quick check 2*

Implantation

The zygote rapidly divides forming a blastocyst which moves to the uterus. If the uterus lining has correctly thickened, on about the 20th day of the cycle, the blastocyst will stimulate the lining to grow around it and the outer layer forms the **trophoblast**. Eventually this forms the **chorion** which develops into the placenta. whilst the inner cells of the blastocyst form the embryo.

Key definition

The blastocyst is the hollow ball of cells that results from the many cell divisions occurring soon after fertilisation. The trophoblast is a layer of cells forming around the outside of the blastocyst. It helps the blastula sink into and attach to the endometrium lining and will later form the chorion.

Hormones in pregnancy

Hormones from the blastocyst are produced as soon as it is implanted. These hormones control the maintenance of the fetus, the birth and lactation.

During pregnancy:
- HCG stimulates the corpus luteum to maintain the pregnancy and the lining of the uterus by secretion of **progesterone** and **oestrogen**.
- Human placental lactogen stimulates the breast tissue to be receptive to oestrogen and progesterone.
- Oestrogen and progesterone maintain the pregnancy and stimulate the breast tissue to develop.
- **Oxytocin** stimulates the uterus to contract and so starts the birth process.

During birth:
- Oxytocin stimulates the contractions to allow birth to proceed. It is a positive feedback control, so as the contractions develop more oxytocin is released.

✓*Quick check 3*

During lactation:
- **Prolactin** promotes and maintains milk production and inhibits ovulation.
- Oxytocin acts as the releaser hormone as the infant suckles and stimulates the muscular contractions around the milk glands that allow milk release.

✓*Quick check 4*

Contraception

There are two strategies used to prevent pregnancy.

Contraception refers to methods of birth control which **prevent fertilisation**:
- birth control pills
- condoms/femidoms
- diaphragm
- injections and implants
- sterilisation.

The other method used **prevents implantation** of the blastocyst:
- intrauterine device (IUD)
- morning-after pill.

Control mechanism	Biological effect	Ethical consideration
Birth control pill	Combination of synthetic oestrogen and progesterone which creates an artificial negative feedback and prevents ovulation	Increased risk of thrombosis and breast cancer increasing the load on medical facilities and society
Condoms/diaphragms/femidom	Acts as a barrier to prevent sperm from meeting the egg	Reduce the risk of sexually transmitted disease, but may be unacceptable from religious viewpoints
Injections and implants, e.g. Norplant® and DMPA (Depo-Provera®)	Contraceptive hormones given subcutaneously that last for up to 3 years	Offer no protection from sexually transmitted disease and carry increased risk of thrombosis and breast cancer which affects society and medical facilities
IUD	Prevents implantation of embryo either physically or by presence of copper	May cause uncomfortable uterine pains and excess bleeding May be considered unethical since the embryo is already formed
The morning-after pill	Used if contraceptive device fails or is not used Large dose of steroids prevents implantation	Removes responsibility and may be considered unethical Causes extreme stomach pain and sickness
Natural rhythm method	Intercourse avoided during ovulation period	None; however not very successful if any factor disrupts the normal menstrual cycle
Sterilisation	The tubes that conduct the sperm from the testis or the ovum from the ovary are cut and tied or clipped to prevent normal passage of the gametes	Individuals may change their minds due to new circumstances, such as new partners or death of a child These operations are not easily reversed

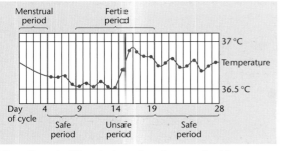

The safe period if ovulation takes place on day 14

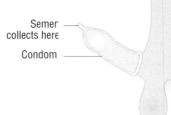

The condom

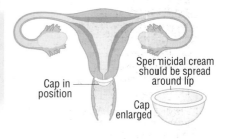

The cap

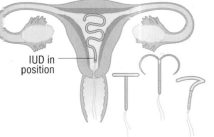

The intrauterine device

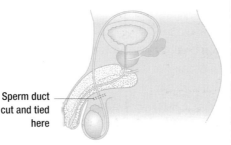

Male and female sterilisation

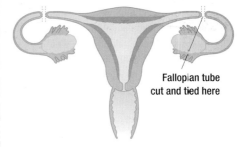

QUICK CHECK QUESTIONS

1. Explain the difference between the blastocyst and trophoblast.
2. What is the function of HCG and how can it be used in pregnancy tests?
3. Explain the term 'positive feedback mechanism'.
4. Describe the roles of oxytocin.
5. Discuss the ethical and religious considerations of the various forms of birth control.

Infertility

Key words

- insemination
- IUI
- ICI
- IVF
- GIFT
- sperm banks
- multiple pregnancy
- monoclonal antibodies
- pre-eclampsia

Examiner tip

Hormone treatments such as follicle-stimulating hormone (FSH) and luteinising hormone (LH) given at the correct times in the cycle have proved successful. Other treatments stimulate the release of gonadotrophin-releasing hormone (GnRH) from the hypothalamus. These all induce ovulation. Check the hormone control of menstrual cycle to understand how these treatments work.

✔ *Quick check 1*

✔ *Quick check 2*

Examiner tip

Learn all the possible fertility treatments for male infertility. Be prepared to use data to comment on the relative success rates.

Examiner tip

You should learn the different types of assistance available for female infertility and understand how they help. Recap your reproduction theory.

Causes of infertility

There are many different causes of infertility.

Females	Males
Abnormal hormone levels so no ovulation	Very low numbers of sperm
Blockage of Fallopian tube	Blockage in sperm duct
Abnormality in uterus lining (e.g. endometriosis)	Abnormal sperm formation
Antibodies may develop that attack the sperm	Produce antibodies that attack their own sperm

Fertility treatments

There are a number of fertility treatments such as using hormones/surgery to unblock Fallopian tubes or sperm ducts, and assisted fertilisation.

Artificial insemination

Semen may be injected into the woman to assist fertilisation.

IUI (intrauterine insemination) involves injecting semen near the Fallopian tube. This requires medical assistance since the tube used to introduce the sperm must pass through the cervix.

With **ICI** (intracervical insemination) the sperm are injected near the cervix in the vagina. This method can be completed at home.

IVF (*in vitro* fertilisation)

Several oocytes and many sperm are mixed together outside the body and kept at body temperature. Some of these may fertilise and form zygotes.

The successful zygotes are allowed to grow into blastocysts for 2–3 days before being transplanted into the woman's uterus.

There are some variations on this technique, such as GIFT (gamete intrafallopian transfer) and ICSI (intracytoplasmic sperm injection).

✔ *Quick check 3*

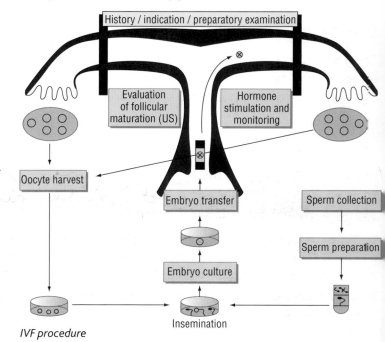

IVF procedure

Sperm bank

Donated sperm is frozen and thawed and used later to fertilise oocytes with a good rate of success. Recent changes in the law may reduce the number of donors, as the identity of the donor can now be revealed if the child applies to the HFEA (Human Fertilisation and Embryology Authority) once they are aged 18 or over.

✔ *Quick check 4*

Embryo storage

After initial screening for diseases human embryos can be frozen aged 1–6 days. They are stored in liquid nitrogen at −200 °C. Correct storage is vital to prevent damage. The embryos may be thawed and implanted when the woman's cycle is at the correct stage. There are strict laws controlling the use of embryos and both partners who contributed the gametes must consent.

Examiner tip

Check the different ways in which embryos may be used and the consent needed.

Multiple pregnancy

This is when more than one fetus develops in the womb simultaneously. Fertility treatment, especially hormone treatments, increases the chances.

✔ *Quick check 5*

Health risks associated with multiple pregnancies include high blood pressure and pre-eclampsia in the mother and low birth weight and premature birth in the babies. Selective reduction in number can be agreed if multiple fetuses result from fertility treatment.

Key definition

Pre-eclampsia is a condition during pregnancy resulting in dangerously high blood pressure and levels of various blood chemicals. The mother and baby are both at risk when she develops pre-eclampsia.

'Vanishing twin' syndrome

A twin observed during ultrasound may disappear later in the pregnancy. The developing fetus has died and, if this is early in the pregnancy, the cells can be absorbed, but later in the pregnancy this can lead to infection and premature labour.

Examiner tip

Most pregnancy tests use **monoclonal antibodies** to detect human chorionic gonadotrophic hormone (HCG) in the woman's urine. You should understand and learn how this procedure works.

Ethical arguments

There are a number of ethical arguments about fertility treatments. When considering these it is important to look at all aspects, including biological and economic arguments, as well as all viewpoints (see page 105).

✔ *Quick check 6*

QUICK CHECK QUESTIONS

1 Explain, using knowledge of control of the ovarian cycle, how hormone treatments may help improve fertility.

2 What is the main difference between IUI and ICI?

3 Describe some of the variations in IVF available and explain how they may assist.

4 Explain the concerns donors may have due to a change in the law regarding donor confidentiality.

5 Give the main risks involved with multiple pregnancies.

6 Describe the steps involved in determining a pregnancy using monoclonal antibodies.

Photosynthesis and respiration

Importance of photosynthesis

Photosynthesis traps and uses light energy to convert carbon dioxide and water into glucose and oxygen. The glucose produced can be used in cellular respiration to produce ATP for active processes such as active transport and protein synthesis. The oxygen released is vital to maintain the oxygen levels in the atmosphere for respiration of all living things.

Photosynthesis forms the basis of many food chains. Producers (photosynthetic plants) can convert light energy into chemical potential energy to pass up the trophic levels in the food chains as organic molecules such as carbohydrates.

Photosynthesis can be divided into two main stages. Both stages occur in chloroplasts found in cells in the green parts of the plant, e.g. palisade mesophyll cell from leaf.

Light-dependent stage

The first stage is the **light-dependent stage** that occurs on the chloroplast membranes called thylakoids. The pigments necessary for trapping the light are found in the membranes grouped into two **photosystems**.

The primary chlorophyll pigment in photosystem 2 absorbs light at a wavelength of 680 nm while that at photosystem 1 absorbs 700 nm.

Light energy excites electrons in the chlorophyll to a higher energy level. They then return to their previous energy level by passing through a series of electron carriers in the membrane, forming ATP and reduced NADP (a coenzyme).

This type of ATP production is known as **photophosphorylation**. Some of the light energy is used to split water – **photolysis** – to provide the source of H^+ and electrons. Oxygen is given off as a waste product.

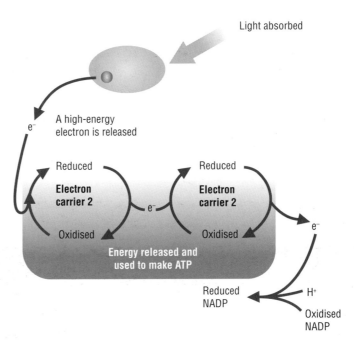

Summary diagrams of light-dependent stage

Light-independent stage

This stage uses the products of the light-dependent stage, ATP and reduced NADP (NADPH), are needed.

This stage is called the **Calvin cycle**. It occurs in the stroma of the chloroplast and each step is controlled by enzymes.

- CO_2 from the atmosphere is combined with 5C **ribulose bisphosphate** (RuBP) using the enzyme **Rubisco** (ribulose bisphosphate carboxylase).

✔️ *Quick check 6*

- An unstable 6C molecule forms and quickly splits into two 3C **glycerate 3-phosphate** (GP) molecules.
- Hydrogen from reduced NADP is used to reduce GP to **triose phosphate** (TP). This also requires some ATP.
- Some TP is used to synthesise glucose and from this other organic molecules.

✔️ *Quick check 4 and 5*

- About 4/5ths of TP is used to regenerate RuBP using more ATP.

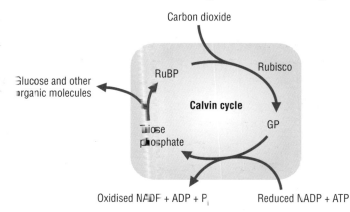

Carbon dioxide
RuBP
Rubisco
Glucose and other organic molecules
Calvin cycle
Triose phosphate
GP
Oxidised NADP + ADP + P_i
Reduced NADP + ATP

Summary diagram of light-independent stage

✔️ *Quick check 2*

The role of CO_2

✔️ *Quick check 6*

CO_2 provides the necessary carbon to build all the organic molecules synthesised by the plant.

TP can be used to synthesise other molecules such as glycerol. The glucose formed may also be built up into sucrose for transport or starch for storage or cellulose for new cell walls.

GP can be directly converted to acetyl co-A which can form fatty acids. These can be added to the glycerol molecules formed to make triglycerides.

If a supply of nitrate is available, amino acids may also be formed from GP or TP. These can be built up into proteins.

The exact products formed depend on the needs of the individual cells or of the plant as a whole.

QUICK CHECK QUESTIONS

1 Explain the link between photosynthesis and respiration.

2 Describe the features of the chloroplast that make it a suitable site for photosynthesis.

3 State the main products of the light-dependent stage.

4 Why is the light-independent stage called a cycle?

5 What is the importance of Rubisco?

6 Explain the importance of CO_2 in photosynthesis and how it is linked to the food chain.

UNIT 4

Cycles

Key words

- nitrogen fixation
- *Nitrosomonas*
- *Nitrobacter*
- *Rhizobium*
- trophic level

Examiner tip

You need to identify some of the microorganisms involved in the N_2 cycle.

Nitrosomonas*, *Nitrobacter and ***Rhizobium*** are the key microbes to remember. Nitrosomonas and Nitrobacter are nitrifying bacteria, Rhizobium is a nitrogen-fixing bacteria.

✓*Quick check 1*

Key definition

Trophic levels are the feeding levels in an ecosystem. Energy is transferred from one level to the next with energy lost at each level as heat.

✓*Quick check 2*

Nitrogen cycle

Nitrogen gas in the atmosphere is not accessible directly by plants.

- A nitrogen source such as ammonium or nitrate ions is essential for plants to manufacture essential amino acids from triose phoshate (TP) and glycerate-3-phosphate (GP).
- The ions are usually available in the soil and are taken up by the plant roots.
- Some plants have nitrogen-fixing bacteria in their roots.
- These microorganisms and other nitrogen-fixing bacteria in the soil convert nitrogen gas into ammonium ions for use by plants.
- The plants use the amino acids they produce to synthesise proteins for new cell material.
- The plant protein is used as a food source for other organisms (herbivores).

The nitrogen is recycled on death as nitrogen-containing compounds are broken down and converted back to nitrate by nitrifying bacteria.

The nitrogen cycle

Energy flow

Energy flows from the light energy converted by green plants in photosynthesis to chemical energy in organic molecules passed on to animals in the food chain.

As the energy is passed from one **trophic level** to the next, the organism uses some of the energy to synthesise new tissue or for other uses such as active transport. Some of the energy is lost as heat or when the organism respires. The result is less energy available for the next trophic level and so fewer organisms can be supported at higher levels in food chains.

A simple food chain

Organism:	maize	→	cattle	→	humans
Energy passed on:	10 kJ		1 kJ		0.05 kJ

Some of the potential energy in the maize (biomass) is not consumed or digested by the cattle, e.g. roots and stems. The energy in these parts is lost to the food chain. Some energy is used by the cattle to form bone, hooves or hide, which are not eaten by humans. In addition, the cattle use up energy in movement, and heat is lost in respiration as well as in excretion. This energy is also lost to the food chain. Decomposers make use of some of the energy lost as organic molecules in dead plant and animal remains or in animal waste.

Each trophic level is therefore inefficient in passing on food energy. Only about 10% of the total energy at one trophic level is passed on to the next.

Carbon cycle

Carbon atoms form approximately 20% of biological molecules.

- Carbon is passed from one organism to another as macromolecules in the food chain.
- It returns to the atmosphere as CO_2 from respiration and decomposition,
- or as CO_2 from fossil fuels when they are burnt (combustion).
- CO_2 is then taken up by plants for photosynthesis. **✓Quick check 5**

✓Quick check 3

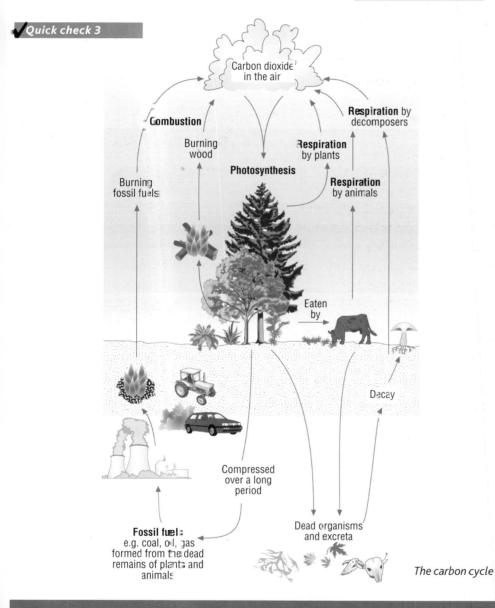

The carbon cycle

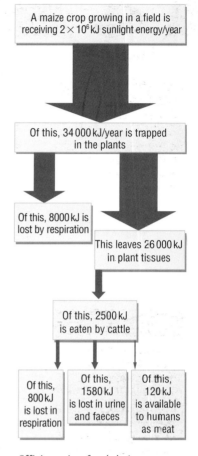

A maize crop growing in a field is receiving 2×10^6 kJ sunlight energy/year

Of this, 34 000 kJ/year is trapped in the plants

Of this, 8000 kJ is lost by respiration

This leaves 26 000 kJ in plant tissues

Of this, 2500 kJ is eaten by cattle

Of this, 800 kJ is lost in respiration

Of this, 1580 kJ is lost in urine and faeces

Of this, 120 kJ is available to humans as meat

Efficiency in a food chain

Examiner tip

Some energy transfer is more inefficient than that; for example, transfer from cow to human is less than 5%. A change in the food farmed and consumed could improve the transfer efficiency. Impala, a grazing animal, transfer as much as 10% of the energy available to them.

✓Quick check 4

QUICK CHECK QUESTIONS

1 Explain the role of microorganisms in the nitrogen cycle and name the key ones involved.

2 Give the reasons for less energy being passed on at each trophic level.

3 Calculate the energy transfer efficiency if cattle consume 2500 kJ, but lose 800 kJ in respiration and 1580 kJ in waste.

4 Why could a change in the type of animals we farm for meat improve efficiency of energy transfer?

5 What are the main processes involved in the carbon cycle?

Food production and the ecosystem

Key words

- extensive
- intensive
- sustainable
- biodiversity
- succession
- ecosystem
- carbon footprint

Extensive farming

This is the term used for farming methods producing crops over large areas but involving little added artificial fertiliser. The natural fertility of the system and recycling of nutrients are relied upon to provide the nitrates for crop growth. This method is only sustainable if overstocking does not occur.

Advantages:
- low maintenance cost
- little environmental damage.

Disadvantages:
- low yield
- limited stock can be supported.

Intensive farming

These are farming methods where large amounts of food are produced over a relatively small area and time. Nutrient loss must be replaced by using artificial fertilisers. In addition, pesticides and herbicides are needed to remove competition by parasites, pathogens and weeds.

✓ Quick check 1 + 2

Advantages:
- high yield which is needed to supply the increasing population
- keeps food cost down.

Disadvantages:
- excessive use of fertiliser and pesticides
- high energy demand for machinery
- reduces biodiversity
- increases environmental damage.

Sustainable farming

This is farming where production can be maintained with little negative effect upon the environment while still providing an income for the farmer.

Agriculture and conservation

Farming affects the ecosystem because **biodiversity** will be reduced since most crops are grown as monocultures. Hedges are removed, land may be burnt and grazing prevents the climax community from developing. Use of fertilisers and pesticides causes damage to surrounding rivers and to the water table.

✓ Quick check 2

Human impact on biodiversity is also caused by factors such as:

- destruction of habitats
- depletion of fish stocks
- extinction of species
- increase in pollution
- competition from non-native species.

Succession

A community of organisms will alter their habitat over time. In turn, this change results in further community changes as some organisms disappear and new ones appear. Gradually over **time** there is a directional change in the composition of the community.

It is difficult to observe succession, but sand dunes provide an opportunity. Near the sea edge, colonisation is beginning by pioneer plants. Further back, the dunes are more stable and a different plant community can be found.

Each stage (sere) results in a distinct community which is replaced when changes lead to the next stage. Factors such as grazing can halt the succession so it is maintained at a stage. This is deflected succession with the stage being referred to as the plagioclimax.

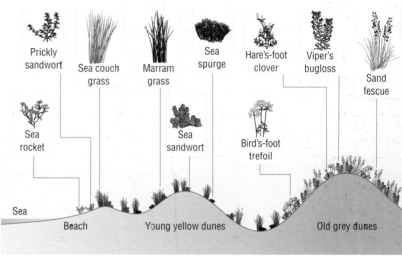

Cross-section of a sand dune showing stages of succession

Ecosystem

An **ecosystem** is a community of living (biotic) organisms interacting in an area and the non-living factors (abiotic) which affect them.

There are three aspects to an ecosystem:

- the habitat where organisms live
- the population of all the organisms of a particular species living together at the same time
- the community of all populations of all different species present.

Human population

The human population has exploded in the last 100 years.

Population increase is determined by surviving birth rate, death rate and migration. Currently the human birth rate is exceeding death rate, creating an exponential rise in numbers.

Globally, infant mortality is lower and rate of recovery from disease has increased due to improvements in healthcare, medicine and technology.

This increase impacts on the environment as more food is required and more waste needs to be removed; more building land and other resources are also needed.

Carbon footprint

This is a measure of our impact on the environment. It is measured as tonnes of CO_2 per year. An increase in atmospheric levels of CO_2 is changing the climate, leading to rising global temperatures, more violent storms and increased flooding.

Key definition

Your carbon footprint is a measure of your annual energy usage, the type of energy used, your petrol use per year and use of air travel.

QUICK CHECK QUESTIONS

1 Compare the advantages and disadvantages of intensive and extensive farming.
2 What environmental damage may be caused by intensive farming?
3 State the stages in sand dune succession.

4 Describe the effects of medical advances, agriculture and disease control on the human population.
5 Discuss the ways that your carbon footprint may be reduced.

End-of-unit questions

1 Respiration is the process by which the energy in food molecules is made available for an organism to do biological work. The energy in the food molecules is transferred to molecules of ATP.
The basic structure of ATP is shown in Figure 1.

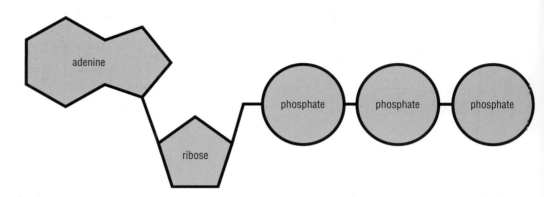

Figure 1

(a) ATP may be described as a phosphorylated nucleotide.
Describe *two* differences in structure between a nucleotide in DNA and ATP. (2)

(b) When ATP in solution is hydrolysed in a test tube, the release of energy quickly heats the surrounding water.

(i) Suggest why this sudden release of heat energy would be dangerous inside a living cell. (2)

(ii) Explain how the energy in ATP is used in order to prevent this happening inside a living cell. (2)

(c) Cyanide is a poison that prevents the production of ATP inside cells. If cyanide is added to cells in a culture before meiosis begins, the nucleus does not divide.

(i) Describe the events occurring inside a cell during prophase I of meiosis. A labelled diagram may be used if it helps to explain your answer. (Synoptic – refers to AS.) (4)

(ii) State *two* uses of ATP during meiosis. (Synoptic – refers to AS.) (2)

2 The fitness of skeletal muscles in an athlete can be improved by a programme of aerobic training.

These muscles are adapted, by their complex structure, to perform well during exercise.

Most skeletal muscles contain fast-twitch white fibres and slow-twitch red fibres. Aerobic training improves the efficiency of respiration in slow-twitch red fibres.

(a) (i) Outline the changes that may occur in **slow-twitch** red fibres during a programme of aerobic training. (4)

(ii) Explain how **fast-twitch** white muscle fibres of an athlete respire during a short burst of strenuous exercise, such as sprinting. (3)

(b) Figure 2 is a diagram of a sarcomere from a skeletal muscle fibre.

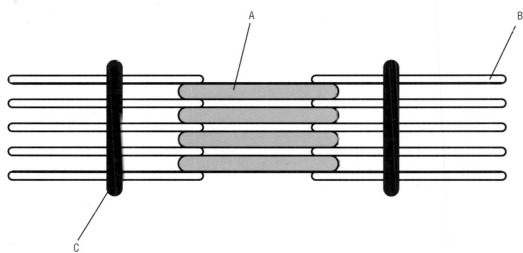

Figure 2

Name **A** to **C**. (3)

(c) Describe the sequence of events occurring in the sarcomere during muscle fibre contraction

In your answer, you should make clear the sequence of events. (Synoptic – this will not appear on A2 papers.) (7)

(d) Suggest how body builders may modify their diet to increase the **development** of skeletal muscle and enhance their performance. (12)

3 An understanding of respiration is important when working out a training programme to improve fitness.

(a) When muscles exercise, the concentration of carbon dioxide and hydrogen ions in the blood increases. These products of respiration are rapidly removed by an increase in the rate and depth of breathing.

Figure 3 is a flow diagram that shows how the rate and depth of breathing is controlled.

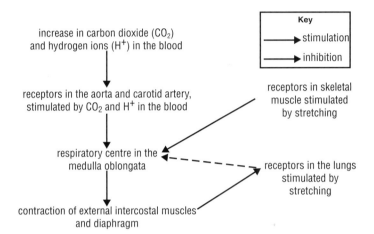

Figure 3

(i) Describe how carbon dioxide and hydrogen ions enter red blood cells from exercising muscle. (3)

(ii) State the enzyme responsible for production of carbon dioxide and the stage of respiration where this enzyme is found. (1)

(iii) Suggest the role of the stretch receptors in the lungs. (2)

(b) Explain why it is so important to control the concentration of hydrogen ions in the blood. (3)

(c) Muscle cells contain large numbers of mitochondria that supply the ATP for muscle contraction. The ATP is produced by oxidative phosphorylation.

Use the most appropriate word or words to complete the paragraph below on oxidative phosphorylation.

Hydrogen atoms from the Krebs cycle are brought to the electron transport chain carried as .... NAD or FAD. Here the hydrogen atoms are split into protons and electrons. The electrons flow through carriers and during this process .... is transferred. The protons are pumped across the inner membrane, producing a proton .... As the protons flood back through the stalked particles, the enzyme .... forms ATP from ADP and inorganic phosphate. (3)

4 Trends in a human population can be monitored by carrying out a census.
Figure 4 shows the age distribution of the populations of the UK and the Philippines
in the year 2000

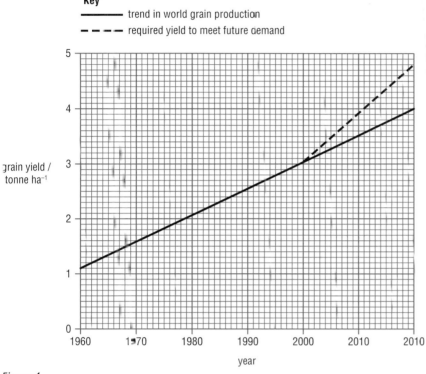

Key
———— trend in world grain production
– – – – required yield to meet future demand

Figure 4

(a) (i) Using the information in Figure 4, state the number of children
aged 0–9 years in the Philippines.   (1)

(ii) The total population of the Philippines in 2000 was 80 million.
Calculate the percentage of the total population of the Philippines who
are aged 0–9 years.
Show your working and express your answer to the nearest
whole number   (2)

(b) Outline reasons for the differences in the **age distributions** of the
populations of the UK and the Philippines.   (6)
In this question one mark is available for linking the explanation to the
pattern shown in Figure 4.   (1)

The human population has continued to grow, largely due to the
development of intensive methods of food production. However, there is a conflict
between some agriculturalists who support intensive methods, and conservationists,
who believe that more food could be produced using extensive methods.

(c) Distinguish between **intensive** and **extensive** methods of food production. (2)

(d) Wheat has been a major food crop for thousands of years.
Selective breeding has led to improvements in wheat varieties.
In 1984, a trial was carried out in which four varieties of wheat, ranging from a very old variety to a modern one, were grown and harvested under identical conditions. Table 1 shows the results of this trial.

Period when variety first introduced	Grain yield/tonnes per hectare	Height of stem/cm	Biomass/tonnes per hectare
19th century	5.05	14.50	15.00
Early 20th century	5.57	134.00	15.41
1950s	6.69	96.00	14.84
1980s	8.05	78.00	15.88

Table 1

Using the information in Table 1, state *two* improvements made in the characteristics of wheat varieties from the 19th century to the 1980s. (2)

5 (a) Figure 5 shows the mean differences in fertility between men and women from 20 to 80 years of age.
Fertility is expressed as a percentage of the maximum fertility achieved at 30 years of age.

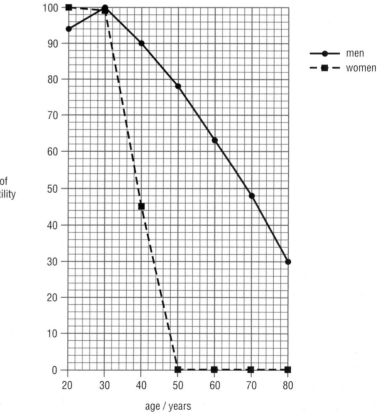

Figure 5

(i) Using the information in Figure 5, describe the changes in fertility of men and women as they grow older. (3)

(ii) Suggest reasons for the difference in fertility between men and women **aged 40**. (3)

(b) (i)   As a woman ages, she is more likely to have twins than a younger
woman.
Suggest a reason for this.   (1)
(ii)   Identical twins are produced when one zygote splits into two, whilst
non-identical twins are produced from two separate zygotes.
Many studies have been conducted comparing non-identical twins with
identical twins. Suggest why these studies are useful.   (3)
(c)   Describe *three* signs or symptoms of the menopause.   (3)
(Synoptic – this will not appear on A2 papers.)

6   There are several different pigments involved in the light-dependent reaction of
photosynthesis in flowering plants.
(a)   Name two photosynthetic pigments found in flowering plants.   (2)

The pigments are arranged in photosystems that absorb light. There are two
photosystems, photosystem I and photosystem II.
(b)   State:
(i)   the precise site where the photosystems are located in chloroplasts   (1)
(ii)   one way in which photosystem I differs from photosystem II.   (1)

Figure 6 is a diagram showing the relationship between the light-dependent reaction
in photosynthesis and the Calvin cycle in a chloroplast.

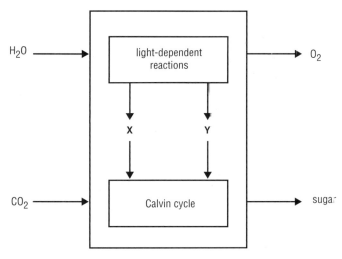

Figure 6

(c)   Name the substances labelled X and Y shown in Figure 6.   (1)
(d)   In this question, one mark is available for quality of spelling, punctuation and
grammar.
In the palisade cells of the leaf the fixation of carbon dioxide is in the Calvin
cycle.
Describe the main features of this cycle. No credit will be given for a flow
diagram of this cycle   (7)
In this question, one mark is available for sequencing the events in the cycle. (1)

Inheritance of human genetic disease

Genetic terminology

Key words

- gene
- allele
- locus
- phenotype
- genotype
- homozygous
- heterozygous
- dominant
- recessive
- codominance
- autosomes
- mutation
- phenylketonuria
- cystic fibrosis
- Huntington's disease
- sickle cell anaemia
- ABO blood groups

Key definition

Genetics is the study of genes, their inheritance and their effects.

✓ *Quick check 1 and 2*

Genetic term	Definition
Gene	A length of DNA which codes for the production of a particular polypeptide
Allele	One of the different forms of a gene which occupy the same locus on homologous chromosomes
Locus	The position on a chromosome at which a particular gene is found
Phenotype	A person's observable characteristics, resulting from an interaction between their genotype and environment
Genotype	The genetic make-up of an organism. It describes all the alleles that the nucleus of a human cell contains. Genotypes can be homozygous or heterozygous
Homozygous	A genotype in which the two alleles of a gene are identical, e.g. AA or aa
Heterozygous	A genotype in which the two alleles of a gene are different, e.g. Aa
Dominant	A dominant allele always shows its effect on the phenotype
Recessive	A recessive allele only shows its effect on the phenotype when the dominant allele is absent
Codominant alleles	Both alleles affect the phenotype in a heterozygote
Autosomes	All the chromosomes except the sex chromosomes (X and Y chromosomes)

A gene mutation in a gamete can be passed from parents to their children. Problems arise when the disease allele is dominant or when the same recessive disease allele is present on both chromosomes in a pair.

✓ *Quick check 3*

Disease	Gene mutation	Result of the mutation
Phenylketonuria (PKU)	A **recessive gene mutation** in a gene on chromosome 12	An error of metabolism caused by the lack of a functional enzyme, phenylalanine hydroxylase. This results in abnormally high levels of the amino acid phenylalanine. A build-up of phenylalanine can lead to severe brain damage in young children
Cystic fibrosis (CF)	A **recessive gene mutation** in a gene on chromosome 7. The mutation is in the gene for cystic fibrosis transmembrane regulator protein (CFTR)	The deletion of three of the base pairs in the gene results in the loss of an amino acid. This makes the cystic fibrosis transmembrane regulator protein (CFTR) defective and triggers the disease process. A result of the defect is to block the movement of chloride ions and water across the membranes of cells in the lungs, liver, pancreas, digestive tract and reproductive tract, and to cause the secretion of abnormally thick, sticky mucus. This mucus obstructs the airways and blocks the secretion of digestive enzymes from the pancreas and the transport of sperm. Daily therapy is needed to help cystic fibrosis patients cough up the mucus ✓ *Quick check 4*
Huntington's disease	An **autosomal dominant gene mutation** (only one allele needs to be defective). The gene is located on chromosome 4	CAG is the genetic code for the amino acid glutamine. The mutation causes the triplet to be repeated so a series of them (i.e. …… CAGCAGCAG ….) produce Huntington proteins with variable numbers of glutamine residues. People with Huntington's disease have approximately 40 or more of these triplets. Disease onset is most common between 30 and 50 years of age and is characterised by progressive mental and physical deterioration ✓ *Quick check 5*

In some cases both alleles affect the phenotype – this is **codominance**. Examples are the inheritance of **sickle cell anaemia** (see page 122) and the **ABO blood groups**.

ABO blood groups are determined by a protein in the cell surface membrane of the red blood cells. There are two forms of this protein – antigen A and antigen B. The gene that codes for this protein has three alleles. The gene locus is represented by I. I^A and I^B are codominant. I^O is recessive to both I^A and I^B.

There are six possible genotypes and four possible phenotypes.

Genotype	Phenotype
$I^A I^A$	Group A
$I^A I^B$	Group AB
$I^A I^C$	Group A
$I^B I^B$	Group B
$I^B I^O$	Group B
$I^O I^O$	Group O

✔ *Quick check 6*

✔ *Quick check 7*

Examiner tip

The symbols for the alleles follow the convention for codominant alleles so, in the case of ABO blood groups, the gene locus is represented by I so the three alleles are; I^A, I^B and I^O

The gene (allele) which causes cystic fibrosis is a recessive gene (allele)
Let C = normal gene (allele)   c = cystic fibrosis gene (allele)

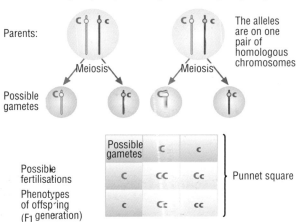

3 normal : 1 with cystic fibrosis

Cystic fibrosis is inherited

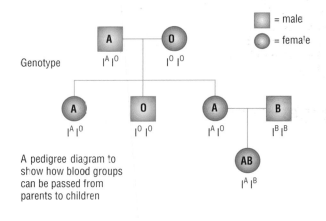

= male
= female

A pedigree diagram to show how blood groups can be passed from parents to children

The inheritance of blood groups

QUICK CHECK QUESTIONS

1 What is an allele?
2 A person's observable characteristics, resulting from an interaction between their genotype and their environment, is called their ....?
3 What is meant by a 'gene mutation'?

4 A mutation in the CFTR gene results in cystic fibrosis. What is the main characteristic of this disease?
5 How does a gene mutation lead to Huntington's disease?
6 What do you understand by the term codominance?
7 What is the genotype for someone with blood group AB?

UNIT 5

Sickle cell anaemia and haemophilia

Key words

- sickle cell anaemia
- codominant
- malaria
- heterozygous advantage
- X and Y chromosomes
- sex linkage
- haemophilia
- factor VIII
- erythrocytes

✓ *Quick check 1 and 2*

Sickle cell anaemia is the result of a mutation in the gene that produces one subunit of haemoglobin. There are two alleles of the gene for the ß-polypeptide chain of haemoglobin; **Hb^A** is the allele for normal ß polypeptide and **Hb^S** the allele for sickle cell ß polypeptide.

There are three phenotypes:

Genotype	Phenotype
Hb^{A}Hb^A	All normal haemoglobin
Hb^{A}Hb^S	Half the haemoglobin is normal and half is sickle cell haemoglobin – sickle cell trait The alleles are **codominant**
Hb^{S}Hb^S	All sickle cell haemoglobin – sickle cell anaemia

Inheritance of sickle cell anaemia

Hb^A – the allele for normal ß polypeptide
Hb^S – the allele for sickle cell ß polypeptide

If the parents are both heterozygous for this trait:

Parental genotypes: Hb^A Hb^S Hb^A Hb^S
Parental gametes: Hb^A Hb^S Hb^A Hb^S

Cross

	Hb^A	Hb^S
Hb^A	Hb^A Hb^S Normal	Hb^A Hb^S Sickle cell trait
Hb^S	Hb^A Hb^S Sickle cell trait	Hb^S Hb^S Sickle cell anaemia

As equal numbers of each type of egg are produced, the chances of each of these four possibilities are equal. The probability of a child being Hb^{A}Hb^A is 0.25, the probability of being Hb^{A}Hb^S is 0.25 and the probability of being Hb^{S}Hb^S is 0.5.

Inheritance of sickle cell anaemia

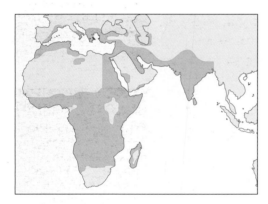

Distribution of malaria

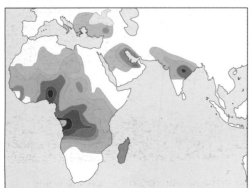

Distribution of sickle cell anaemia

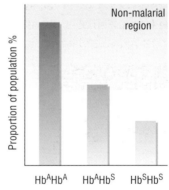

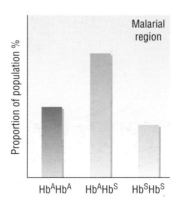

Distribution of sickle cell anaemia in a malarial and non-malarial population

Sickle cell anaemia and malaria

Malaria is one of the most common infectious diseases. It is caused by *Plasmodium*, a parasite belonging to the Protoctista (single-celled eukaryotic organisms).

Children with sickle cell trait (Hb^{A}Hb^S) have shown resistance to the malaria parasite, with the level of resistance increasing with age. The mutant allele is therefore advantageous and is selected for. This results in a causal link between the parts of the world where sickle cell anaemia is most common and the parts of the world where malaria is endemic.

Possible explanations

1 It may be that Hb^AHb^S causes the malaria infection to stay in the body for longer (the bulk of blood cells carrying malaria are destroyed quickly but a few may escape detection), so allowing the immune system to build up a proper defence against malaria.
2 During the stage of its life cycle when the parasite lives inside red blood cells its metabolism changes the internal chemistry of the red blood cell. The haemoglobin is likely to become deformed and be destroyed before the daughter parasites emerge – fewer parasites means less chance of serious malaria infection.

This means that having the genotype Hb^AHb^A is a **disadvantage** in the areas where malaria is **endemic**, and a child with genotype Hb^SHb^S is unlikely to survive to reproduce, so the best genotype to have in such areas is Hb^AHb^S. In each generation, children born with the genotype Hb^AHb^S – **heterozygous advantage** – are most likely to grow up and reproduce, so both alleles are passed on to the next generation increasing the frequency of the mutant allele.

✓ *Quick check 3*

Sex linkage and disease

Women have two X chromosomes, while men have one X and one Y. In humans, the Y chromosome is much shorter than the X. The X chromosome carries a large number of genes whereas the Y chromosome carries very few. The result is that some sex-linked characteristics in men (XY) are controlled by only one allele on the X chromosome with no matching allele on the Y. Any gene that is carried on either the X **or** Y chromosome is said to be **sex-linked**. There are very few genes on the Y chromosome.

One example of a sex-linked disease is **haemophilia**.

The **dominant allele, H**, codes for the production of a protein called **factor VIII**, whilst the **recessive allele, h**, results in a lack of the factor. Females who are homozygous recessive or males who have only the one recessive allele will have haemophilia.

Genotype	Phenotype
X^HX^H	Normal blood clotting female
X^HX^h	Normal blood clotting female but a carrier for haemophilia
X^hX^h	Haemophiliac female very unlikely
X^HY	Normal blood clotting male
X^hY	Haemophiliac male

✓ *Quick check 6*

QUICK CHECK QUESTIONS

1 What is the genotype for someone with sickle cell anaemia?
2 Hb^A and Hb^S alleles are codominant. What does this mean?
3 Why is it an advantage to have the sickle cell trait if you live in an area where malaria is endemic?
4 How could you test a person for sickle cell anaemia?
5 Which of the sex chromosomes in humans carries fewer alleles?
6 Draw a genetic cross to show the possible offspring from a female haemophiliac carrier and a normal male.

The use of pedigree diagrams

The **pedigree diagram** is a pictorial description of a family tree. The first step is to collect information – a complete record includes information from three generations of relatives, including children, brothers and sisters, parents, aunts and uncles, nieces and nephews, grandparents, and cousins.

Using the pedigree diagram it is possible to predict the risks of a particular couple having a child who might suffer from a particular genetic disease. Pedigrees are often used to determine the mode of inheritance (dominant, recessive, etc.).

For example, a woman may have a brother who has **haemophilia**, which would mean that her mother and her grandmother must have been carriers. If her mother is a carrier then there is a 50% chance that she is also a carrier and 25% chance that she will give birth to a haemophiliac son. A famous pedigree chart shows the inheritance of haemophilia from Queen Victoria in members of various European royal families.

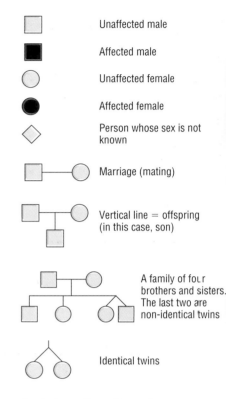

Symbols used in pedigree charts

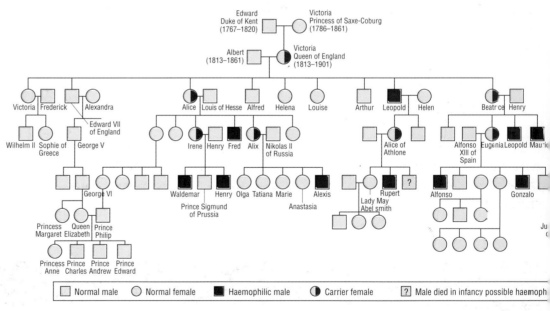

A pedigree chart showing the distribution of haemophilia in the European royal families

Autosomal linkage

Linkage is when genes that are close together on the same chromosome tend to stay together during the formation of gametes.

An example of this is the inheritance of the ABO blood groups together with nail patella syndrome.

This syndrome affects males and females equally so is not sex-linked. It is autosomal dominant and the locus of the gene that causes it is found on chromosome 9 very close to the ABO blood group gene. These genes will tend to be inherited together, as the chance of a **chiasma** forming to separate them is unlikely.

The use of crossover frequencies

Crossovers occur during prophase 1 when the chromatids of a bivalent may break and reconnect to another chromatid. This results in the exchange of alleles between maternal and paternal chromatids. The new genotypes are called **recombinants**.

The closer together two gene loci are on a chromosome, the less likely a crossover is to occur between them. The farther they are apart, the more likely a crossover is to occur between them.

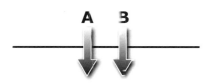

The alleles of genes A and B are far apart on this chromosome so crossovers are likely

The alleles of genes A and B are very close together on this chromosome so crossovers are unlikely

Linkage mapping shows the positions of gene loci along a chromosome. A unit of measure is necessary:

1 map unit = 1% crossing over

The frequency of recombinants is the crossover value:

$$\frac{\text{number of recombinant individuals produced}}{\text{total number of offspring}} \times 100$$

It is therefore possible to map the relative positions of gene loci using the data of crossover frequencies; two genes whose alleles give a crossover value of 20% are said to be 20 units apart. In this way, we can map the linear sequence of gene loci along a chromosome.

QUICK CHECK QUESTIONS

1 What is the use of a pedigree diagram?

2 Looking at the pedigree diagram of the European royal family:
  (a) How many haemophiliac males are shown?
  (b) Which of Queen Victoria and Prince Albert's children were carriers of haemophilia?
  (c) From the pedigree chart is haemophilia sex-linked and is it recessive or dominant?

3 What is meant by the term autosomal linkage?

4 What are crossover frequencies?

5 Two gene loci A and C show a 30% crossover value. Suppose that gene locus C has a 9% crossover with B and 21% with A. Draw a line to show where you would expect A, B and C to occur on a chromosome.

Non-disjunction and translocation

Key words

- non-disjunction
- translocation
- Turner's syndrome
- Klinefelter's syndrome
- Down's syndrome
- trisomy
- mosaic
- karyotype

✓ *Quick check 1, 2 and 3*

Examiner tip

Review the formation of gametes.

These are both examples of a chromosome mutation.

- In **non-disjunction** there is a change in the number of chromosomes.
- In **translocation** a piece of chromosome breaks off and is transferred to another chromosome.

Non-disjunction is the failure of two members of a homologous pair of chromosomes to separate properly during meiosis. Non-disjunction of the sex chromosomes during meiosis can result in either a gamete carrying two sex chromosomes or no sex chromosome, instead of a single one. Examples are Turner's and Klinefelter's syndromes. In **Turner's syndrome** all or part of one of the X chromosomes is missing (X0). Turner's syndrome is the most common sex chromosome abnormality of human females.

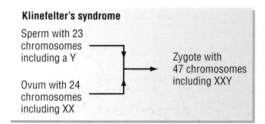

Klinefelter's syndrome

Sperm with 23 chromosomes including a Y

Ovum with 24 chromosomes including XX

Zygote with 47 chromosomes including XXY

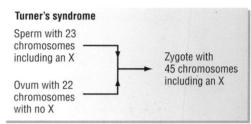

Turner's syndrome

Sperm with 23 chromosomes including an X

Ovum with 22 chromosomes with no X

Zygote with 45 chromosomes including an X

How Turner's syndrome and Klinefelter's syndrome occur

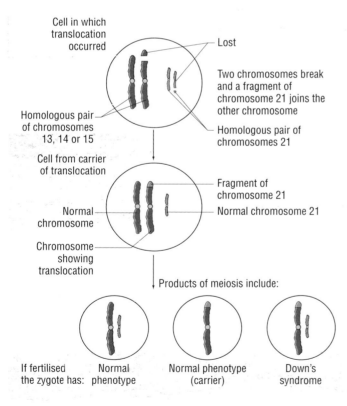

Cell in which translocation occurred

Lost

Two chromosomes break and a fragment of chromosome 21 joins the other chromosome

Homologous pair of chromosomes 13, 14 or 15

Homologous pair of chromosomes 21

Cell from carrier of translocation

Fragment of chromosome 21

Normal chromosome

Normal chromosome 21

Chromosome showing translocation

Products of meiosis include:

If fertilised the zygote has: Normal phenotype

Normal phenotype (carrier)

Down's syndrome

Down's syndrome caused by translocation in chromosome 21

In **Klinefelter's syndrome** affected individuals have at least two X chromosomes and at least one Y chromosome (XXY).

Another disease caused by non-disjunction is **Down's syndrome**.

- **Regular trisomy 21** – all the cells have an extra chromosome 21. Around 94% of people with Down's syndrome have this type.
- **Mosaic** – only some of the cells have an extra chromosome 21. Around 2% of people with Down's syndrome have this type, which tends to result in milder features. Mosaics can also occur with Turner's syndrome.

Down's syndrome can also occur as the result of **translocation** when the end of the long arm of chromosome 21 joins another chromosome. This accounts for around 4% of cases. This type of Down's syndrome can be inherited.

✓ *Quick check 4, 5 and 6*

Confirmation of a diagnosis of Turner's, Klinefelter's or Down's syndrome is by **karyotype analysis**

A **karyotype** is an organised profile of a person's chromosomes. In a karyotype, chromosomes are arranged and numbered by size, from largest to smallest. It is then possible to see if there are any chromosome abnormalities.

Ethical issues around Down's syndrome, Turner's syndrome and Klinefelter's syndrome are concerned with attitudes, feelings and preferences about how we ought to live, and what decisions we ought to make.

Medical research seems to be concentrating its resources not on preventing babies from being born with a disability, but on preventing babies with a disability from being born, which is quite different.

Examiner tip

Make sure you know the related AS work.

	Down's syndrome	Turner's syndrome	Klinefelter's syndrome
Effects	Most common cause of mental retardation and malformation in a newborn	Ovaries don't function properly so she may not develop secondary sexual characteristics and may be infertile. May have learning difficulties and other medical problems	Most common symptom is infertility. Other symptoms: tall, enlarged breasts, small penis, lack of facial hair. These can lead to bullying in school
Is it common?	Occurs in about 1 in every 800–1000 births	Affects approximately 1 in 2500 girls	Found in about 1 out of every 500–1000 newborn males
Treatment?	No cure, but treatment of health problems and support for learning difficulties allows many to lead relatively normal and semi-independent lives. Others need full-time care	No cure; emotional support is vital (may have low self-esteem). Growth hormone and oestrogen treatment to achieve development of secondary sex characteristics and adult height	No cure; speech and physical therapy, testosterone treatment

Some points to consider:

- When it is discovered prenatally that a woman is carrying a child with one of the above syndromes, should she abort the pregnancy?

- Despite the wide availability of prenatal genetic screening and facilities for termination, the rate at which babies who have Down's syndrome are being born is rising in England and Wales.

- Many fetuses that do not have Down's syndrome are lost following tests such as amniocentesis.

- Support for the families involved needs to be 'ongoing'. Cost considerations?

Examiner tip

When discussing ethical issues try to marshal your ideas and keep to the point.

QUICK CHECK QUESTIONS

1 A change in chromosome numbers occurs in ....?

2 What happens in translocation?

3 What chromosome is missing in Turner's syndrome?

4 Klinefelter's syndrome only occurs in ....?

5 What is the cause of Down's syndrome?

6 If a karyotype showed three copies of chromosome 21 what would the person be suffering from?

Gene technology

Key words

- restriction enzymes
- blunt ends
- sticky ends
- recombinant DNA
- anneal
- ligase
- intron
- exon
- palindromic
- gel electrophoresis
- polymerase chain reaction
- DNA profiling
- minisatellites

Hint

A gene consists of **introns** – regions that are not translated into proteins – and **exons** – the regions that are translated (see page 94).

Key definition

Annealing – this is when the sticky ends, formed when two pieces of DNA have been cut with the same enzyme, stick to each other by forming hydrogen bonds to complementary base pairs.

Ligase is an enzyme that catalyses the condensation of two molecules.

Hint

Restriction enzymes are named after the bacteria species they came from. Two examples of restriction enzymes are *Eco*R1, which comes from *Escherichia coli* strain R and was the first to be identified, and *Hind*lll is the third enzyme from *Haemophillis influenzae*.

✔*Quick check 3 and 4*

In genetic engineering **restriction enzyme**s are used to cut up and join together parts of the DNA of one organism, and insert them into the DNA of another organism. In the resulting new organism the inserted genes will code for one or more new characteristics.

✔*Quick check 1*

- The first stage is to identify, isolate and extract the required gene.

Restriction enzymes (restriction endonucleases) will produce a double-stranded cut in the DNA at a particular sequence of bases (the target site). Some restriction enzymes cut straight across both chains of DNA, forming **blunt ends**, but the most useful enzymes make a staggered cut in the two strands, forming **sticky ends**. Restriction enzymes are highly specific – they will only cut DNA at a specific base sequences. There are thousands of different restriction enzymes known.

✔*Quick check 2*

- The second stage is putting the isolated gene into another organism forming **recombinant DNA**.

Once the section of DNA has been cut out and **annealed** with the DNA of another organism the broken DNA needs to be repaired. **DNA ligase** repairs the DNA backbone by forming covalent phosphodiester bonds; it links up the sugar-phosphate backbones of the newly paired section.

Restriction enzyme forming sticky ends

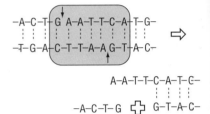

Bases T-T-A-A left exposed so these will readily join with their complementary bases of A-A-T-T

Palindromic sticky ends

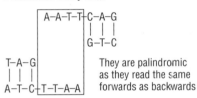

They are palindromic as they read the same forwards as backwards

Sticky ends and palindromes

Recognition sites are palindromic

Many restriction enzymes are **palindromic** – the cuts correspond to nitrogenous base sequences that read the same backwards and forwards. Two types of DNA sequences are possible.

- Mirror-like palindrome – a sequence that reads the same forwards and backwards on the same DNA strand, as in GTAATG.

- Inverted repeat palindrome – here the forward and backward sequences are found in complementary DNA strands so GTATAC is complementary to CATATG. This is more common.

Gel electrophoresis is a form of chromatography used to separate different fragments of DNA on the basis of their length. Details of the process are shown in the figure on the next page.

The DNA fragments are invisible at this stage. They can be stained with azure A (blue strands) or ethidium bromide, a fluorescent dye. A more sensitive way would be to label the DNA samples at the start with a radioactive isotope such as ^{32}P. Photographic film is placed on top of the finished gel in the dark, and the DNA shows up as dark bands on the film. The fragments may be lifted from the gel onto special paper for further analysis. This technique is called Southern blotting.

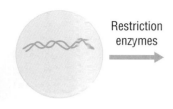

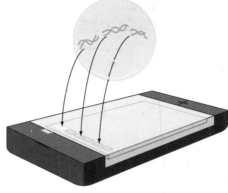

Restriction enzymes

1 Restriction enzymes cleave DNA into smaller segments of various sizes

2 DNA segments are loaded into wells in a porous gel. The gel is made of agarose. The gel floats in a buffer solution within a chamber between two electrodes. The buffer solution enables the electrical current to flow through the gel

3 When an electric current is passed through the chamber, DNA fragments move toward the positively-charged anode. DNA is negatively charged because of the phosphates that form the backbone of the DNA molecule

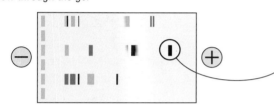

4 Smaller DNA segments move faster and farther than larger DNA segments

Gel electrophoresis

For further information on introns and exons see page 94.

Polymerase chain reaction (PCR) is a technique of replicating DNA in a test tube. PCR can clone DNA samples as small as a single molecule. It depends on the fact that the DNA bases are complementary.

- Denaturation – high temperatures cause the hydrogen bonds holding together the double helix to be disrupted. The molecule unzips or 'denatures'.

- Annealing – the DNA solutions are allowed to cool and mixed with primers. Primers bind and form small sections of double-stranded DNA.

- Elongation – DNA polymerase adds free nucleotides to the unwound DNA extending the double-stranded section. The DNA will be replicated as many times as the cycle is set for.

✓ *Quick check 5*

DNA profiling involves identifying the patterns of genetic material. Genetic fingerprinting and profiling exploit the highly variable repeating sequences (minisatellites). Two unrelated humans will be unlikely to have the same number of **minisatellites** at a given locus. In DNA profiling a small sample of human tissue, even from partially degraded samples, is taken and analysed using electrophoresis to obtain a banding pattern. Banding patterns can then be stored on a database.

> **Key definition**
>
> Primers are short, single-stranded sequences of DNA, around 10–20 bases in length. They are needed in PCR to bind to a section of DNA, because DNA polymerase enzymes cannot bind directly to single-stranded DNA fragments.

> **Hint**
>
> DNA profiling is used in forensic science, paternity testing and matching organ donors.

> **Key definition**
>
> A minisatellite is a section of DNA that consists of a short series of bases, 10–100 bp; these occur at more than 1000 locations in the human genome.

QUICK CHECK QUESTIONS

1 What is meant by a restriction enzyme?
2 Distinguish between blunt and sticky ends on a piece of DNA.
3 How does electrophoresis sort fragments of DNA by length?

4 Why do DNA fragments move to the anode in gel electrophoresis?
5 What are the main stages in PCR?

Genetic engineering in microorganisms

Key words

- vector
- plasmid
- recombinant DNA
- gene therapy
- augmentation
- germ cells
- somatic cells

Hint

The technology of using antibiotic-resistant genes has been superseded by the use of 'reporter' systems (using an enzyme which produces fluorescent products with appropriate substrates).

Hint

Human insulin is produced using genetically engineered bacteria.

Examiner tip

Bacteria cannot be used to make all human protein because in eukaryotic cells genes need to be 'edited' after transcription and the polypeptide may need modifying after translation in the Golgi apparatus. These processes would not necessarily happen in prokaryotic cells so a functional protein would not be made.

Hint

Review your knowledge of transcription and the role of the Golgi apparatus.

✔ *Quick check 2 and 3*

1 Identify the required gene.
2 Use restriction endonucleases to cut the gene out of the donor organism.
3 Prepare the **vector** by cutting the bacterial plasmid using restriction enzymes which leave sticky ends corresponding to those of the donor gene.
4 Mix the gene and vector (the **plasmid**).
5 The new gene incorporates into the bacterial plasmid. The sticky ends are lined up and the gene is attached – annealed – then DNA ligases join the pieces of DNA together to make a recombinant plasmid (a plasmid containing **recombinant DNA**).
6 Incorporate the engineered plasmid into the bacterium (transformation). Once the plasmid is inside the host bacterium, the genetically modified bacterial cells will multiply.
7 Only a small proportion of the cells will take up the recombinant DNA so marker genes, such as an antibiotic resistant gene, are inserted.

Key definitions

A vector is used to transport the DNA into the host cell.

A bacterial plasmid is a small circular strand of DNA often found in bacteria in addition to their main DNA. Plasmids are often used as vectors.

✔ *Quick check 1*

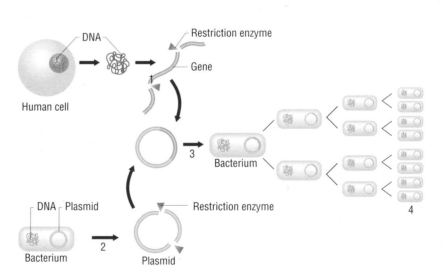

Genetic engineering in microorganisms

Genetic engineering using a eukaryotic cell

Bacteria cannot always make human proteins. In animals the only secretions that can be collected are milk and urine.

Examples of human protein made in this way are anti-thrombin, an anti-blood clotting agent used in surgery, and alpha 1-antitrypsin (AAT), an enzyme used to treat cystic fibrosis and emphysema.

The technique of using a eukaryotic cell line to produce a human protein can be understood by looking at how AAT is produced in the following diagram.

Gene therapy

Gene therapy is an experimental technique that modifies genes to treat or prevent disease. This is done by:

- adding a gene (copy of a functional allele) – this is called **augmentation**
- repairing an abnormal gene
- altering the degree to which a gene is turned on or off.

✔ *Quick check 4*

Gene therapy has the potential to correct the underlying cause of a disease. Many diseases such as cystic fibrosis, muscular dystrophy, haemophilia and cancer are caused by faulty genes. Gene therapy involves the addition of a healthy, working copy of the faulty gene into the appropriate cells of the body.

The two main challenges are getting the gene into the cells more efficiently and to make gene expression last longer. Problems include the body's immune response destroying the vector and the fact that many disorders, such as heart disease, Alzheimer's disease and diabetes, are caused by the combined effects of variations in many genes.

Ethical implications of gene therapy

The practice of gene therapy relates to two groups of cells.

1 **Germ cells** – gene therapy using cells which give rise to eggs or sperm results in permanent changes that are passed down to subsequent generations. The appeal of germ-line gene therapy is the possibility of eliminating some diseases from a particular family, and ultimately from the population. However, this also raises controversy as it could open the door to the irreversible alteration of the human species and the genetic change started may actually be deleterious and harmful.

2 **Somatic cells** are non-reproductive, so this therapy is viewed as a safer approach, because only the targeted cells in the patient are affected, and the allele is not passed on to future generations.

✔ *Quick check 5 and 6*

A female sheep is given a fertility drug to stimulate her egg production, and several mature eggs are collected from her ovaries	
The eggs are fertilised *in vitro*	
A plasmid is prepared containing the gene for human AAT and the promoter sequence for b-lactoglobulin. Hundreds of copies of this plasmid are microinjected into the nucleus of the fertilised zygotes. Only a few of the zygotes will be transformed, but at this stage you cannot tell which	
The zygotes divide *in vitro* until the embryos are at the 16-cell stage	
The 16-cell embryos are implanted into the uterus of surrogate mother ewes. Only a few implantations result in a successful pregnancy	
Test all the offspring from the surrogate mothers for AAT production in their milk. This is the only way to find if the zygote took up the AAT gene so that it can be expressed. About 1 in 20 eggs are successful	
Collect milk from the transgenic sheep for the rest of their lives. Their milk contains about 35 g of AAT per litre of milk. Also breed from them in order to build up a herd of transgenic sheep	
Purify the AAT, which is worth about £50 000 per mg	

How AAT is produced

Examiner tip

Review your knowledge of the immune response.

Hint

Human germ cell therapy is illegal in the UK.

QUICK CHECK QUESTIONS

1 What is the role of a vector in genetic engineering?

2 Why can't microorganisms be used to make some of the human proteins?

3 How is the genetically engineered human protein collected from animals?

4 Explain what is meant by gene therapy.

5 Which two groups of cells are involved in gene therapy?

6 What are the main concerns about using gene therapy on germ cells?

Human Genome Project and genetic counselling

Key words

- Human Genome Project
- pedigree analysis
- genetic counselling

Human Genome Project

This project was devised to map and sequence the entire human genome; that is, to locate every gene on every human chromosome and to sequence all the base pairs of coding and non-coding regions. In April 2003, researchers successfully completed the Human Genome Project.

- Genomes are first mapped to identify which chromosome they have come from.
- Samples of the genome are sheared into smaller sections.
- These sections are placed into separate bacterial artificial chromosomes (BACs) and are transferred to *E. coli* cells.
- The *E. coli* cells grow in culture producing clones of the sections.
- DNA is extracted from these cells and restriction enzymes used to cut it into smaller fragments (different restriction enzymes giving different fragment types).
- Fragments are separated by electrophoresis.
- Each fragment is sequenced using an automated process.
- Computer programs analyse the overlaps shown by fragment sequences.

✓ *Quick check 1*

Possible uses of this knowledge

- Maps generated by project researchers have helped in finding genes associated with dozens of genetic conditions, Huntington's disease, myotonic dystrophy, fragile X syndrome, etc.
- There are now more than 1000 genetic tests for human conditions. These tests enable patients to understand their genetic risks for disease and also help healthcare professionals diagnose disease.
- The ability to sequence any person's genome may start an age of personalised medicine where an individual's genetic code can be used to prevent, diagnose and treat diseases.

✓ *Quick check 2*

How pedigree charts can indicate the probability of genetic disease occurring

Pedigree charts are a pictorial description of a family tree. Using **pedigree analysis** it is possible to predict the risks of a particular couple having a child who might suffer from a particular genetic disease. Pedigrees are often also used to determine the mode of inheritance (dominant, recessive, etc.). (See also page 124.)

✓ *Quick check 3*

The role of the genetic counsellor

Genetic counsellors work with people concerned about the risk of an inherited disease. These could be:

- new parents or couples planning a pregnancy
- couples with a baby or child who has a physical problem or shows delay in development
- couples who have lost a baby during pregnancy or infancy
- family members concerned that they too may carry an allele for a disorder
- couples with a known genetic condition in either family
- couples who are close blood relatives.

Genetic counsellors explain the risks of passing on a genetic disorder to the next generation and/or what the choices are. This enables a person to make an informed decision about the options available to them, and to understand the nature of the disease and what having it will mean in practical terms. Implications for other family members are also explored.

Ethical issues involved in the work of the genetic counsellor

One of the main ethical dilemmas arises from a conflict between the right of the individual to personal privacy and the interest of other family members.

- In the case of planning a pregnancy when genetic risks are high, those involved are left with three options: (i) prenatal diagnosis and abortion if required, (ii) artificial insemination (sperm can be screened, or donor sperm used), and (iii) gene therapy. Should the genetically defective be aborted? Do parents have a right to produce defective children?

- Genetic counsellors learn many family 'secrets', such as previous abortions, previous abnormal births and, occasionally, false paternity. The father may believe that he is a carrier, but then the test on him proves negative. Should the family be protected from the disruption resulting from disclosure?

> **Examiner tip**
>
> Outline ethical issues clearly. Avoid emotive phrases such as 'playing God' and keep to the facts. Consider the possible consequences to individuals, to families and to society, as well as the rights of the unborn child.

✔ *Quick check 4*

QUICK CHECK QUESTIONS

1 When sequencing human DNA how are the fragments separated?

2 What are the possible uses of mapping the human genome?

3 Using the pedigree chart: If Leon and Jane have another child, what are the chances that it will have Duchenne muscular dystrophy?

4 What is the main role of a genetic counsellor?

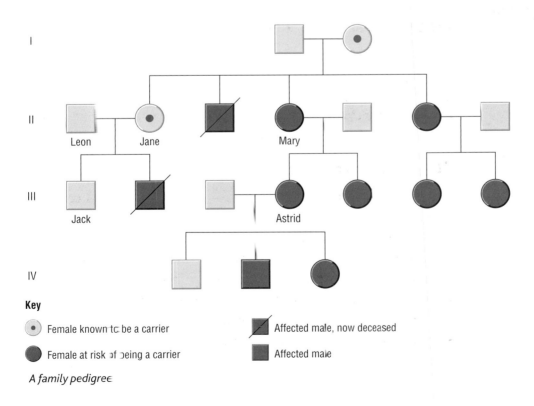

Key

⊙ Female known to be a carrier

⬤ Female at risk of being a carrier

▨ Affected male, now deceased

■ Affected male

A family pedigree

UNIT 5

Transplant surgery and cloning

Key words

- transplant surgery
- HLA antigens
- donor
- recipient
- tissue rejection
- compatibility
- haplotype
- major histocompatibility complex
- xenotransplantation
- stem cells
- pluripotent
- blastocysts

Key definition

HLA stands for human leucocyte antigen.

Key definition

When **donor** tissue is transplanted into a **recipient**, the recipient will produce an immune response against the transplanted tissue. This is known as **tissue rejection** as the tissue will eventually be destroyed. **Compatibility** is the ability to accept transplanted tissue.

✔ *Quick check 1*

Key definition

The area of chromosome 6 where the loci of the genes for HLA antigens are situated is called the **major histocompatibility complex (MHC)**.

✔ *Quick check 2*

✔ *Quick check 4*

In **transplant surgery** it is difficult to find a suitable donor because the donor's and patient's 'tissue type' must closely match. A person's tissue type is defined by genetic markers on the surface of cells called **HLA antigens**. These genetic markers are inherited, so best matches are identical twins. Siblings are much more likely to have similar HLA antigens than unrelated people.

If a donor must be located in the general population, the chances of finding a match range from 1 in 1000 to one in several million.

Genes determining HLA antigens are found on chromosome 6 and there are six gene loci involved, designated HLA – A, B and C, and HLA – DR, DQ and DP. Each of the HLA antigens occurs, in different individuals, in as many as 20 varieties, so that the number of possible HLA types reaches about 10 000. Each set of HLA antigens is referred to as a **haplotype** and every person inherits one haplotype from each parent.

Sources of donated organs

Source	Advantages	Disadvantages
Cadavers	Cannot harm the dead by removing an organ	Number of available organs has fallen (better road safety and medical advances) Getting consent from grieving relatives
Living donors	Better outcome than those from a cadaver Reduced waiting time	Risk to donor
Organ sale	On demand??	Fear of exploitation
Xenotransplantation (animal-to-human transplants)	Shortage of donor organs	Immunological rejection The risk of catching a disease communicable from animals to humans, particular concern about retroviruses

✔ *Quick check 3*

Potential of genetic engineering in the use of non-human organs for transplant surgery

In the future, whole organs or tissue from pigs may be transplanted into human recipients. A pig is the optimal donor because:

- their organs are a comparable size to humans
- they are easy to breed.

However, pig tissue has a number of specific proteins and carbohydrates on the cell surface membranes which trigger a very strong human immune response. Genetic modification has the potential to modify these such that they become compatible.

Potential for cloning human embryos to create a supply of embryonic stem cells

Stem cells are **pluripotent**. Cultured stem cells will make exact copies of themselves until different chemicals, or molecular signals, trigger them to differentiate.

Stem cells can be obtained by:

- using **blastocysts** that are left over from *in vitro* fertilisation
- cloning embryos and harvesting stem cells from them, the embryos are destroyed before they are 14 days old
- taking them from aborted fetuses. ✔*Quick check 5*

Why are embryonic stem cells needed?

- Stem cells offer the hope of a renewable source of replacement cells and tissues.
- They can be used to test new drugs.
- They can be used to make what are often called 'designer babies'. ✔*Quick check 6*

Ethical issues

Therapeutic and reproductive cloning	Transplant surgery
<ul><li>Could it lead to a new eugenics when babies are designed to suit social preferences?</li><li>Possibly only the rich will have access to the technology – one class with a superior genetic make-up, other class the undercogs?</li><li>What percentage of human genes does an organism have to contain before it is considered human?</li><li>Is stem cell research for the benefit of human health ethically justifiable?</li><li>If therapeutic cloning became common practice, large numbers of human embryos would be created, and once the stem cells had been removed, destroyed</li></ul>	<ul><li>Should we be able to stipulate who our organs go to?</li><li>The shortage of available kidneys has led some to suggest using prisoners, street children and brain-damaged patients as organ donors</li><li>Should we be able to opt in or opt out of organ donation?</li><li>What about paying for an organ?</li><li>Compensation for time off work for living donors?</li></ul>

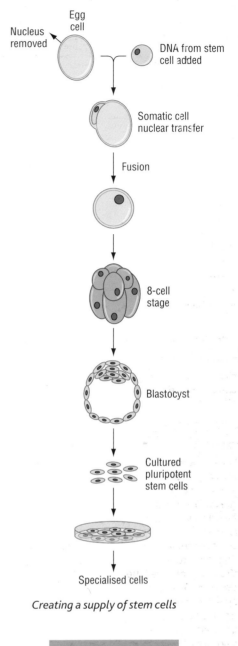

Creating a supply of stem cells

> **Key definition**
>
> Pluripotent cells have the potential to develop into many different cell types

QUICK CHECK QUESTIONS

1. What do you understand by the term 'genetic compatibility' in transplant surgery?

2. What is the major histocompatibility complex?

3. What are the two main sources of donated organs in the UK?

4. What are the main problems of using pigs for human organ transplants?

5. From where can stem cells be obtained?

6. State three uses of embryonic stem cells.

135

UNIT 5

Visual function

Key words

- nervous system
- central nervous system
- peripheral nervous system
- sclera
- choroid layer
- retina
- fovea centralis
- conjunctiva
- rods
- cones
- cornea
- iris
- pupil
- ciliary body
- lens
- suspensory ligaments
- aqueous and vitreous humours
- optic nerve
- bipolar cells
- ganglion cells

✔ *Quick check 1*

✔ *Quick check 2, 3 and 4*

The function of the **nervous system** is communication. The **central nervous system** (CNS) consists of the brain and spinal cord. The **peripheral nervous system** connects all parts of the body to the CNS. The peripheral (sensory) nervous system receives stimuli, the CNS interprets them, and then the peripheral (motor) nervous system initiates responses.

Structure of the eye

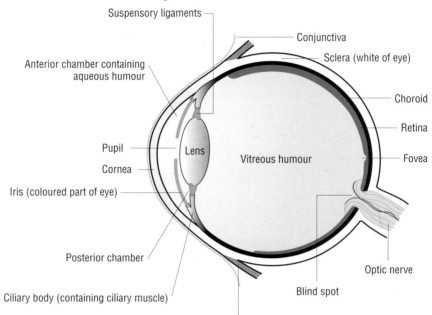

The structure of the human eye

Name of structure	Structure	Function
Sclera	A tough, white outer layer	Protects structures within it and maintains the shape of the eye
Choroid layer	Richly supplied with blood vessels; inner part made up of cells containing the dark pigment, melanin	Absorbs light and prevents it from being reflected inside the eye
Retina	Innermost layer of the eye. Contains the receptor cells: **rods** and **cones**	Receptor cells receive the light stimulus and convert light into nerve impulses
Fovea	Part of the retina where cones are concentrated	Area of maximum visual acuity
Conjunctiva	Thin layer at the front of the eye	Protects the surface of the eye. Kept moist by fluid secreted by the tear ducts
Cornea	Continuous with the sclera	Refracts light rays entering the eye so focusing light rays onto the retina
Iris	Contains circular and radial muscles and pigmented cells	Helps to control the amount of light passing into the eye
Pupil	Circular space (hole) in the centre of iris	Size altered by the contraction and relaxation of the iris muscles
Ciliary body	Contains ciliary muscles	Helps to control the shape (diameter) of the lens
Lens	Made up of stacks of long, narrow, transparent cells; about 4 mm thick and biconvex	Focuses rays of light onto the retina
Suspensory ligaments	Ligaments that run between the lens and the ciliary body	Hold the lens in place. Accommodation
Vitreous humour	A gelatinous fluid found behind the lens	Maintains the shape of the eye
Aqueous humour	A watery liquid found in front of the lens	Maintains shape of front of eye
Optic nerve	Bundle of nerve fibres	Carries action potentials to the brain

Structure of the retina

Light energy is converted in the retina to nerve impulses (action potentials) that are carried along the optic nerve to the brain.

There are two types of light-sensitive receptor cells – **rods and cones**. These are outermost in the retina whereas **bipolar** and **ganglion** cells lie closest to the lens. Light therefore travels through the thickness of the retina before striking and activating the rods and cones.

Rod and cone cells have very similar structures.

Rod cells	Cone cells	Common to both
Elongated structure	Shorter, broader and more tapered than rods	Inner segment contains the nucleus and numerous mitochondria
Outer segment contains many discs of membrane-enclosed sacks. The membrane contains the pigment	No discs – the membranes (lamellae) are formed from one continuous folded surface	Outer segment specialised for photoreception
Only one type of pigment visual purple, called **rhodopsin**	Three forms of the pigment iodopsin (each most sensitive to a different wavelength of light). Once pigment is bleached by light, it takes about 6 min to regenerate	End of inner segment forms synapses with other cells in retina
Able to work in low light intensity because can respond to a single photon of light. Do not detect colour		

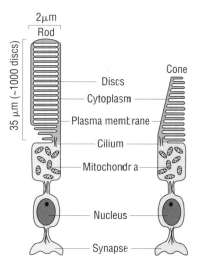

The structure of the light receptor cells

- The **fovea centralis**, at the centre of the macula, contains only cones.
- The **blind spot** is where the optic nerve leaves the retina. There are no rods or cones here.
- **Bipolar cells** have a central body and two sets of processes. Those nearest the rods and cones are short, and branch into many endings forming synapses with either a number of rods or a single cone. The other process is longer, and forms synapses with a ganglion cell.
- **Ganglion cells** have numerous dendrites which form synapses with the bipolar cells. It is here that action potentials are first generated in the retina.

✓*Quick check 5*

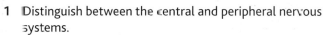

QUICK CHECK QUESTIONS

1　Distinguish between the central and peripheral nervous systems.

2　Which structure in the eye has the function of controlling the amount of light that enters the eye?

3　What is the function of the lens?

4　Which receptor cells are adapted to detect colours, and function well in bright light?

5　Where are action potentials first generated in the retina?

The response of sensory receptors to a light stimulus

Action potentials

The **action potentials** are transmitted to the brain along nerve cells in the optic nerve.

A **resting potential** of −40 mV is maintained across a rod cell's surface membrane.

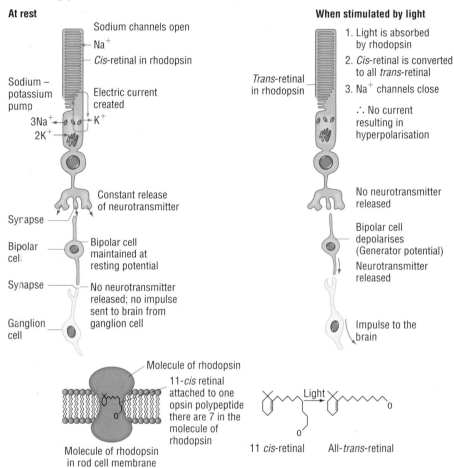

At rest

Sodium channels open
Na⁺
Cis-retinal in rhodopsin
Sodium – potassium pump
Electric current created
3Na⁺
K⁺
2K⁺
Constant release of neurotransmitter
Synapse
Bipolar cell maintained at resting potential
Synapse
No neurotransmitter released; no impulse sent to brain from ganglion cell
Ganglion cell

When stimulated by light

1. Light is absorbed by rhodopsin
2. *Cis*-retinal is converted to all *trans*-retinal
3. Na⁺ channels close

∴ No current resulting in hyperpolarisation

Trans-retinal in rhodopsin

No neurotransmitter released

Bipolar cell depolarises (Generator potential)

Neurotransmitter released

Impulse to the brain

Molecule of rhodopsin
11-*cis* retinal attached to one opsin polypeptide there are 7 in the molecule of rhodopsin
Molecule of rhodopsin in rod cell membrane

Light
11 *cis*-retinal   All-*trans*-retinal

How light generates an action potential

Summary

✓ *Quick check 1 and 2*

Rod cells in the dark	Rod cells in the light
Opsin + *cis*-retinal → **rhodopsin** This uses energy in the form of ATP	**Rhodopsin** → opsin + *trans*-retinal (the rhodopsin absorbs light and changes shape; 11 *cis*-retinal has a kink at carbon 11; *trans*-retinal has no kink)
Sodium ion channels open	Sodium ion channels close
Membrane depolarised	Membrane hyperpolarised
Neurotransmitter (glutamate) released from rod cell to bipolar cell	No neurotransmitter released into synapse
Bipolar neurone hyperpolarised	Bipolar neurone depolarised
No neurotransmitter released into synapse between bipolar cell and ganglion	Neurotransmitter released into synapse between bipolar cell and ganglion
No generator potential	Generator potentials add together to form action potential
No action potential	Action potential travels across synapse to ganglion cell and from there along the optic nerve to the brain

The change in shape of the retinal molecule and the consequent change in shape of the opsin protein is called **bleaching**.

Using routine eye tests to assess receptor activity

Visual acuity – acuteness or clearness of vision.

This is measured by using a **Snellen chart**. This chart has a series of letters or letters and numbers, with the largest at the top. You would be asked to read down the chart.

The smaller the letter you can read at a fixed distance the better your visual acuity. If you can resolve letters approximately one inch high at 20 feet you are said to have 20/20 visual acuity.

The results of this test will tell the optometrist what kind of lenses you need to help you see more clearly.

A Snellen chart

Colour vision

About 8% of males and 0.5% of females have some version of 'colour blindness' from birth. This is usually a genetically inherited sex-linked trait, and is of the 'red-green confusion' variety. Total colour blindness (seeing in only shades of grey) is extremely rare.

It is tested by looking at a collection of cards made up of different coloured spots which are arranged so that those of a particular colour form images or numbers.

Pupil response test

You sit in a dimly illuminated room and a bright light is shone in your eye by swinging a torch so that light shines first into one eye and then the other. Light shone into one eye should make both pupils constrict equally. If there is a difference then the person is said to have a relative afferent papillary effect (RAPD). This could indicate damage to the optic nerve or the brain. It can also indicate that the person's nervous system is affected by alcohol or other drugs. This test is also suitable for babies and unconscious patients.

How the blink response can be used to indicate levels of consciousness

The **pupil response test** relies on a reflex action and it does not require the patient to communicate.

The **blink reflex** is one of the last to be lost as unconsciousness deepens; its absence indicates that the person is in a coma. An eye opening indicates that the arousal mechanism in the brain is active.

✓ *Quick check 6*

QUICK CHECK QUESTIONS

1 In the light what changes occur in the pigment rhodopsin?
2 Is the bipolar neurone depolarised or hyperpolarised in the light?
3 What type of chart is used to test for visual acuity?

4 How can colour vision be tested?
5 What does the pupil response test indicate?
6 What is the blink reflex?

The brain and autonomic nervous system

The gross structure of the human brain

The **brain** is protected by the **cranium**.

Inside the cranium, the brain is surrounded by the **meninges** (connective tissue); these membranes separate the skull from the brain.

The cells in the blood vessel walls are joined tightly, forming the **blood-brain barrier.**

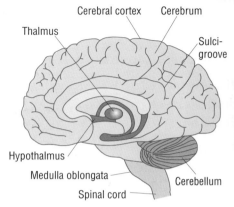

The structure of the human brain

Key definition

Blood-brain barrier: a mechanism that alters the permeability of brain capillaries, so that some substances, such as certain drugs, are prevented from entering brain tissue, while other substances are allowed to enter freely (e.g. oxygen).

The brain contains ventricles filled with **cerebrospinal fluid** (cerebral fluid) which is continuous with the fluid in the spinal cord.

The shape of the ventricles is quite distinctive and can be seen clearly on a magnetic resonance imaging scan.

The meninges help to secrete **cerebral fluid** which helps to protect and cushion the brain – it allows the brain to 'float'.

The brain can be divided into three parts.

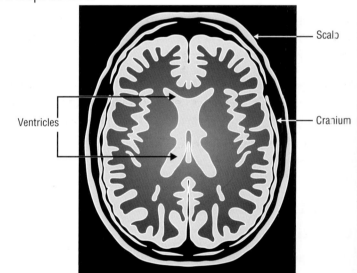

Part of the brain	Position and gross structure
Forebrain or **cerebrum**	The largest part of the brain. It is made up of two cerebral hemispheres connected to each other by a bridge of tissue called the **corpus callosum**. Highly folded tissue called the cerebral cortex covers the surface of the cerebral hemispheres. The folds make the brain more efficient (larger surface area so more neurones). Each hemisphere can be divided into four lobes: frontal, parietal, temporal and occipital
Midbrain	Located in front of the cerebellum. The cerebrum, the cerebellum and the spinal cord are all connected to it
Hindbrain or **cerebellum**	Behind the cerebrum at the back of the head. Contains over half of all the nerve cells in the brain
Medulla oblongata	Lies beneath the cerebellum. Forms the link between the brain and the spinal cord

Functions of parts of the brain

Part of the brain	Function
Cerebrum	• Frontal lobe – contains the centres for speech, memory, intelligence and emotion. It also initiates voluntary motor activity. There is a special region on the left side, called **Broca's area**, which helps control the patterns of muscle contraction necessary for speech • Parietal lobe (behind the frontal lobe) – general sensation and the integration of senses • Occipital lobe (at the centre back of the cerebrum) – vision • Temporal lobe (at either side of the cerebrum) – involved in hearing. **Wernicke's area** has to do with understanding language. The **hippocampus**, found deep in the temporal lobe, plays an important role in learning and memory formation
Cerebellum	Involved in coordinating your muscles to allow precise movements and control of balance and posture

Medulla oblongata	Controls autonomic functions and relays nerve signals between the brain and spinal cord
	Examples of processes controlled include:
	    Respiration – the rate and depth of breathing
	    Blood pressure, heart rate, reflex arcs

✓ *Quick check 4, 5 and 6*

The **autonomic nervous system** operates largely independently of conscious control. It carries nerve impulses from the central nervous system to effectors such as the smooth muscle of the gut, blood vessels and heart, and also to glands. There are two branches. They use different neurotransmitters and so have different, often antagonistic, effects.

Examiner tip

Link the functions listed above with your knowledge of movement, vision, respiration, etc.

- **Sympathetic**: generally acts to arouse the body, preparing it for 'fight or flight'. Most active in times of stress, e.g. increases heart and ventilation rate, dilates pupil.

- **Parasympathetic**: acts to relax and restore the body to normal levels of arousal. Most active in sleep and relaxation, e.g. decreases heart and ventilation rate, constricts pupils.

Examiner tip

Sympathetic and parasympathetic stimulation occurs continuously.

✓ *Quick check 7*

QUICK CHECK QUESTIONS

1 What is the function of the cerebral fluid?

2 Which part of the brain forms a link between the brain and the spinal cord?

3 Which is the largest part of the brain?

4 Which area of the brain is connected with understanding language?

5 What is the function of the cerebrum?

6 Where would you find the part of the brain that controls the rate and depth of breathing?

7 Distinguish between the peripheral and the autonomic nervous systems.

Neurones are specialised cells that are adapted to transmit nerve impulses very rapidly from one part of the body to another. There are three types of neurones: **sensory**, **motor** and **relay**.

Neurones consist of:

- a **cell body** – this contains the nucleus and large amounts of rough endoplasmic reticulum and its associated ribosomes, a well developed Golgi apparatus and large numbers of mitochondria.

- **axon** – this is a long process that carries the impulse away from the cell body.

- **dendrites** – very short processes that carry nerve impulses towards the cell body.

Many axons and **dendrons** are surrounded by **Schwann cells**. These wrap themselves many times around the axon building up layers of membrane. The membranes are rich in myelin (a lipid) and so form a **myelin sheath** around the axon. The function of this sheath is to act as an insulator and prevent exchange of ions. The myelin sheath speeds up the transmission of a nerve impulse. The gaps between adjacent Schwann cells are called **nodes of Ranvier** (no myelin sheath here). In humans they occur about every 1–3 mm, and are the only regions where ions can enter or leave the axon.

✔ *Quick check 1*

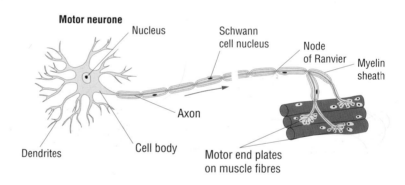

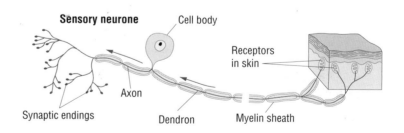

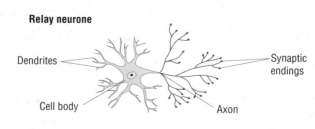

Types of neurones

✔ *Quick check 2 and 3*

	Motor neurone (efferent neurone)	Sensory neurone (afferent neurone)	Relay neurone
Function: to transmit impulses	From the CNS to an effector (muscle or gland)	From a receptor to the CNS	Between neurones, e.g. from a sensory neurone to a motor neurone in the spinal cord
Specific structure	Cell body lies within the brain or the spinal cord. They have a long axon and many short dendrites	One dendron brings the impulse towards the cell body, and an axon carries it away	Numerous short processes

Examiner tip

Motor neurones occur in the autonomic and somatic nervous systems. The effectors for the somatic system will be skeletal muscles: check from page 141 what type of muscle will be the effectors in the autonomic system.

✔ *Quick check 5*

Hint

The simple reflex arc is present from birth so it does not have to be learned.

Key definition

Dorsal root ganglion: a nodule on the dorsal root of the spinal cord that contains cell bodies of sensory neurones.

✔ *Quick check 6*

A **simple reflex** is rapid and entirely automatic. Examples include the sudden withdrawal of a hand in response to a painful stimulus. It is a means of protecting the body from dangerous stimuli.

The response is rapid because:

- the decision-making processes of the brain are not involved
- the neurone pathway is short.

Sensory cells (**receptors**) send impulses to the spinal cord along a sensory neurone. The impulse passes through the dorsal root ganglion into the grey matter of the spinal cord. From here it may pass to the cell body of a motor neurone, or to a relay neurone and then onto a motor neurone. The impulse travels along the motor neurone to an **effector** such as a muscle or a gland.

Within the spinal cord, the impulse will also be passed on to other neurones that take the impulse up the spinal cord to the brain so the brain is 'aware' of the reflex arc.

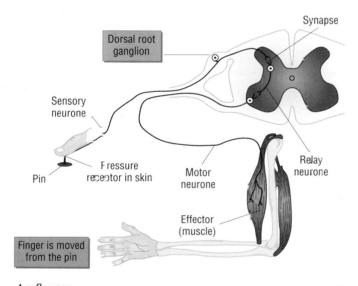

A reflex arc

QUICK CHECK QUESTIONS

1. Where would you find the myelin sheath and what is its function?

2. What is the difference in function between a sensory and a motor neurone?

3. What is the function of a relay neurone?

4. Distinguish between an axon and a dendron.

5. What makes the simple reflex arc a fast reaction to a stimulus?

6. With a line drawing, show the passage of an impulse through a simple reflex arc.

Transmission of a nerve impulse

Each nerve impulse is a brief change in the potential difference (p.d.) across the plasma membrane of the neurone. In nerve and muscle cells the membranes are electrically excitable, this means that their membrane potentials can change. This is the basis of a nerve impulse. When a neurone is not sending a signal, it is 'at rest' – **the resting potential**. The brief change in polarity that is the nerve impulse is called the **action potential**.

The resting potential

- The membrane of a resting neurone is **polarised** – there is a difference in electrical charge (**potential difference**) between the outside and inside of the membrane. The inside is negative (about –70 millivolts) compared with the outside.
- The s**odium/potassium pump** uses ATP to actively pump three sodium ions out of the cell for every two potassium ions it brings into the cell.
- K$^+$ diffuses back out again much faster than Na$^+$ diffuses back in, giving an overall excess of positive ions outside the membrane compared with inside.

The action potential

When a stimulus above a certain intensity or **threshold** arrives at a receptor or nerve ending, its energy causes a temporary and rapid reversal of the polarity of the membrane of the neurone.

- This stimulus causes the voltage-gated Na$^+$ channel to open. Since there is a high concentration of Na$^+$ outside, Na$^+$ diffuses rapidly into the neurone and the cell becomes positively charged inside compared with the outside. The cell is **depolarised.**
- The potential difference across the plasma membrane reaches +40 mV.
- The sodium ion channels close and the voltage-gated potassium ion channels open allowing potassium ions to diffuse out of the cell. The cell is repolarised.
- The p.d. overshoots slightly (–70 to –90 mV) making the cell **hyperpolarised**.
- The resting membrane potential is restored by the Na$^+$/K$^+$ pump as excess ions are subsequently pumped in/out of the neurone and most of the potassium ion channels close.

This transient switch in membrane potential is the **action potential**. This cycle of depolarisation and repolarisation is extremely rapid.

The action potential travels (is propagated) along an axon automatically.
- Local circuits are set up between the depolarised region and the regions on either side of it.
- Sodium ions flow sideways inside the axon, away from the positively charged region towards the negatively charged regions on either side.
- The axon membrane ahead depolarises, generating an action potential.

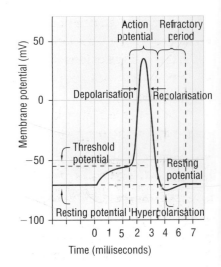

The action potential

'All or nothing' law

The action potential only occurs if the stimulus causes enough sodium ions to enter the cell to change the membrane potential to a certain **threshold** level. If this is not reached then no action potential (and hence no impulse) will be produced. This is called **the all or nothing law**. The ion channels are either open or closed; there is no halfway position.

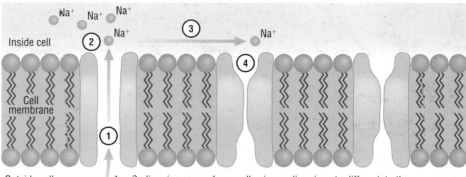

Inside cell

Cell membrane

Outside cell

Na⁺

1 – Sodium ion channel open allowing sodium ions to diffuse into the neurone
2 – Localised increase in concentration of sodium ions inside neurone, the action potential
3 – Sodium ions diffuse along the axon/dendron, away from the region of higher concentration
4 – Sodium gate, which was initially closed, will open because of movement of sodium ions – allowing the action potential to move along the neurone as more sodium ions enter and set up another action potential

The creation of local currents

Refractory period

During this time (about 0.001s), the Na^+ and K^+ channels cannot be opened by a stimulus. This means that **action potentials only move in one direction** along the axon.

✓ *Quick check 5*

Saltatory conduction

In myelinated neurones the voltage-gated ion channels are found only at the nodes of Ranvier. Between the nodes, the myelin sheath acts as a good electrical insulator preventing ions moving in or out. The action potential therefore jumps large distances from node to node (1mm). This is called **saltatory conduction**.

Key definition

Saltatory conduction: wherever the Schwann cells wrap around the axon, sodium and potassium ions cannot cross the membrane. Therefore, the only place that an action potential can occur is at the node of Ranvier. When the action potential is present at one node, the influx of Na^+ ions causes the displacement of K^+ ions down the axon. This makes the next node more positive and depolarises it to threshold. Thus, the action potential jumps from node to node, from *saltare* meaning to jump.

The speed of an action potential is affected by the following.

- The myelin sheath – saltatory conduction. Maximum speed of a nerve impulse in a non-myelinated neurone is around $1\,ms^{-1}$ and $100\,ms^{-1}$ in myelinated neurones.
- Axon diameter – the larger the axon, the faster it conducts. This is because there will be less leakage of ions.
- Temperature – affects the rate of diffusion of ions. The higher the temperature the faster the impulse.
- The number of synapses – the greater the number of synapses in a neurone pathway, the slower the conduction velocity.

✓ *Quick check 6*

Hint

Action potentials are always the same size; however, the **frequency of the impulse** can be determined by the intensity of the stimulus. A strong stimulus produces a rapid succession of action potentials; a weak stimulus results in fewer action potentials per second. Also, a strong stimulus is likely to stimulate more neurones than a weak stimulus.

Hint

A good website for students would be http://www.biologymad.com/NervousSystem/nerveimpulses.htm

QUICK CHECK QUESTIONS

1 What is a resting potential?
2 Distinguish between depolarisation and hyperpolarisation.
3 What is a threshold potential?
4 Explain briefly how an action potential moves along a neurone.

5 What is the importance of the refractory period?
6 What do you understand by saltatory conduction and why is it important?

Synapses and brain injury

A **synapse** is a junction between one neurone and another neurone or effector. A neurone may have up to 10 000 synapses with 1000 other cells. ✓*Quick check 1*

The gap between two cells is known as the **synaptic cleft**.

Action potentials cannot cross a synapse. A **neurotransmitter** carries the transmission across the synaptic cleft. **Cholinergic synapses** are those in which the neurotransmitter is acetylcholine (ACh). ✓*Quick check 2 and 3*

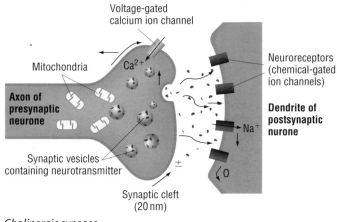

Cholinergic synapse

Synapses consist of:

- **synaptic bulb** – where neurotransmitters are made. In the synaptic bulb are numerous **mitochondria**, together with large amounts of endoplasmic reticulum and synaptic vesicles containing the neurotransmitter
- **postsynaptic membrane** – has receptors for binding the neurotransmitter
- **synaptic cleft** or gap between the presynaptic and postsynaptic membranes.

Transmission across a cholinergic synapse

- An action potential arrives at the membrane of the synaptic bulb.
- The voltage-gated **calcium ion channels** open causing calcium ions (Ca^{2+}) to diffuse into the bulb.
- This causes the **synaptic vesicles**, containing ACh, to fuse with the **presynaptic membrane** and release their contents by exocytosis into the synaptic cleft.
- ACh diffuses across the synaptic cleft and binds to receptors on the postsynaptic membrane. The membrane protein receptors have a complementary shape to ACh.
- This changes the shape of the protein, opening sodium ion channels in the membrane.
- Sodium ions enter the postsynaptic neurone, depolarising the membrane and initiating an action potential, if the threshold is reached.
- The enzyme **acetylcholinesterase** hydrolyses the neurotransmitter **ACh**. The breakdown products are absorbed by the presynaptic neurone by endocytosis and used to synthesise more ACh. This stops ion channels being permanently open on the postsynaptic membrane. ✓*Quick check 5*

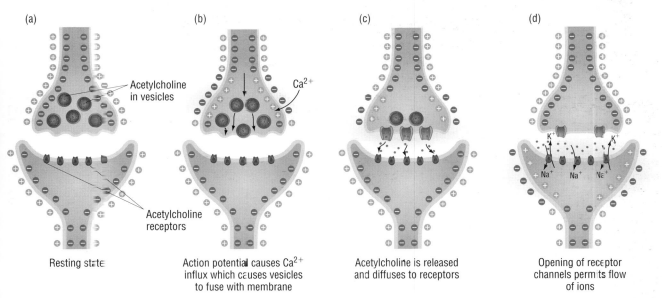

(a) Resting state

Acetylcholine in vesicles

Acetylcholine receptors

(b) Action potential causes Ca^{2+} influx which causes vesicles to fuse with membrane

Ca^{2+}

(c) Acetylcholine is released and diffuses to receptors

(d) Opening of receptor channels permits flow of ions

K^+ K^+

Na^+ Na^+ Na^+

Cholinergic synapse.

The role of synapses

- Synapes allow transmission in **one direction only**, from presynaptic to postsynaptic membrane. This is because synaptic vesicles are confined to the presynaptic side of the cleft, and protein receptors to the postsynaptic side.   ✔ *Quick check 6*

- They provide the means through which the nervous system connects to and controls the other systems of the body such as muscles.

- They provide a means of interconnecting nerve pathways. One presynaptic neurone might diverge to several postsynaptic neurones. The establishment of new synaptic links is the basis of learning.

- They allow the 'filtering out' of continual unnecessary or unimportant background stimuli. If a neurone is constantly stimulated (e.g. clothes touching the skin), the synapse will not be able to renew its supply of transmitter fast enough to continue passing the impulse across the cleft. This 'fatigue' places an upper limit on the frequency of depolarisation.

A **traumatic brain injury** (TBI) is an injury caused by the head being hit by something or shaken violently. The damage may be caused by the head forcefully hitting an object such as the dashboard of a car or by a gunshot wound. The physical, behavioural or mental changes that may result from head trauma depend on the areas of the brain that are injured.

The term TBI is not used for a person who is born with a brain injury or whose brain injuries happened during birth.   ✔ *Quick check 7*

QUICK CHECK QUESTIONS

1   What is meant by the term 'synapse'?

2   What is a neurotransmitter?

3   Why are some synapses called 'cholinergic synapses'?

4   Why does the synaptic bulb contain many mitochondria?

5   List the sequence of events as a nerve impulse is transmitted across a cholinergic synapse.

6   How do synapses ensure that an impulse is transmitted in one direction only?

7   What is traumatic brain injury?

Brain and spinal cord damage

Assessing brain and spinal cord damage

CT (computerised tomography) scan – an X-ray machine linked to a computer takes a series of detailed pictures that build up into a three-dimensional picture. These can show tumours in the brain, narrowing of the spinal canal, and blood clot or intracranial bleeding in patients with stroke or brain damage from head injury.

The **MRI (magnetic resonance imaging) scan** offers the most sensitive non-invasive way of forming a detailed image of the brain and the spinal cord.

The MRI scanner is a tube surrounded by a giant circular magnet. It uses a powerful magnetic field instead of radiation so the scans may be carried out on pregnant women.

An MRI scan shows

- trauma to the brain shown as bleeding or swelling
- brain aneurysms, strokes, tumours of the brain or spinal cord, inflammation of the spinal cord
- demyelination in the CNS, so a powerful tool in the diagnosis of multiple sclerosis
- an increase of water content in the cells of brain tissue (one effect of a stroke); MRI is better than CT in detecting small infarcts soon after stroke onset.
- the type, position and size of the tumour in the brain or spinal cord – enables doctors to plan treatment.

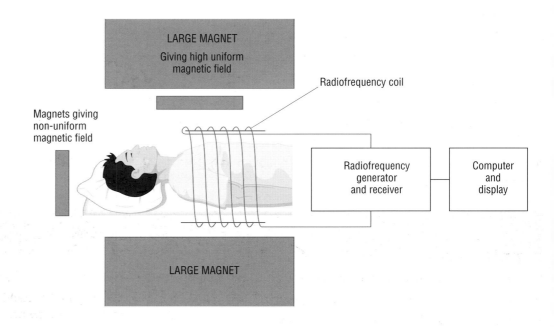

Nerve conduction velocity (NCV) test – both sensory and motor nerves are repeatedly stimulated by electrodes in order to measure the speed of conduction of nerve impulses. Unusually slow conduction velocities suggest damage to nerve fibres (e.g. demyelination). Patients referred for NCV tests suffer from nerve conditions that produce numbness, tingling, pain or loss of sensation.

✔*Quick check 1*

Why are damaged neurones unable to regenerate?

Axon regeneration in the CNS fails because:

- astrocytes, a type of **glial cell**, enlarge and proliferate to form a scar and produce myelin and inhibitory molecules that inhibit the regrowth of a neurone
- most CNS axons only show a feeble regeneration response after they are cut

✔*Quick check 2*

- fibrinogen, a blood-clotting protein found in circulating blood, has been found to inhibit the growth of CNS neuronal cells.

The potential of stem cell research for treating diseases of the nervous system is great, as so many diseases (Parkinson's, Alzheimer's, stroke, etc.) result from the loss of nerve cells.

Successes

- If stem cells are added to spinal implants made of hydrogels, then the hydrogels will help fill the cavities in injured areas and create an environment in which neurones can grow.
- A stroke leaves a permanent gap in the brain. It should be possible to use stem cells to fill the gap.

Stroke can have a detrimental effect on **short-term memory** (a working memory which stores information about recent happenings). **Long-term memory** (the store of autobiographical information, language, acquired knowledge) is usually well preserved. After a stroke the patient will show a noticeable loss of concentration; they may tend to get things out of sequence, or misinterpret or confuse information. They will probably need to have things repeated over and over.

People **caring for a stroke patient** can help them regain some of their short-term memory by:

- showing the patient photographs
  - to help them identify people they used to know
  - in newspapers to help them identify well-known people
- playing a memory game with the patient
- having a large clear calendar in every room so that the patient can keep track of the day
- leaving clear reminder notes around the house so that the patient knows what they should be doing and when.

Examiner tip

Remind yourself about stem cells – this is covered on page 135. Terms like pluripotent and differentiation can apply to the development of nerve cells as well as blood cells.

✔*Quick check 3*

✔*Quick check 4*

Hint

11 February 2008: Researchers at the University of California discover a gene involved in generating the cerebral cortex. The finding could one day lead to stem cell therapies to treat brain injuries, stroke and Alzheimer's.

✔*Quick check 5*

QUICK CHECK QUESTIONS

1 State how brain and spinal cord damage can be assessed.

2 Why are damaged neurones unable to regenerate?

3 Stem cells are pluripotent; what does this mean?

4 What are the effects of a stroke on memory?

5 State two techniques that could be used to help improve the memory in stroke patients.

Drugs

- drug
- legal drugs
- illegal drugs
- Parkinson's disease
- dopamine
- diamorphine
- opioids
- endorphins
- cannabis

Key definition

Drug abuse is the habitual misuse of a chemical substance including illegal drugs, prescription drugs and over-the-counter drugs. 'Abuse' refers to use that is problematic or harmful, either for the individual or those around them.

Hint

Caffeine will keep us awake when we need it, but as the effects persist over 3 hours it will also delay desired sleep.

Hint

People with Parkinson's disease can't simply take dopamine tablets or vegetable products containing dopamine (e.g. fava beans), because dopamine taken by mouth does not get into the brain.

Type of drug	Examples
Medicines	Paracetamol, penicillin, insulin
Everyday substances	Caffeine, nicotine, alcohol
Illegal substances	Cannabis, heroin, cocaine

A **drug** is any chemical that affects the way your body functions, and how you think and feel.

✔*Quick check 1*

Legal drugs, such as alcohol, tobacco and caffeine, and prescription medicines also carry risks of dependency and damage, and can be 'misused' in the same way as **illegal drugs**.

Parkinson's disease is caused by the progressive loss of neurones in a part of the brain that produces the chemical **dopamine**.

Symptoms of Parkinson's disease include tremors, stiffness, slow movements, and impaired balance and coordination.

The enzyme in the synapse that breaks down the dopamine continues to act, so there will not be enough dopamine to balance the normal levels of acetylcholine (ACh). The overall effect is extensive loss of dopamine in the brain.

Most drug treatments for Parkinson's increase the level of dopamine in the brain or oppose the action of ACh.

✔*Quick check 2*

Drug	What it does
Dopamine replacement drugs such as levodopa and a chemical	Breaks down in the body to form dopamine; the chemical ensures the right amount of dopamine in the brain
Drugs that mimic the action of dopamine	May be taken alone before using levodopa to reduce its long-term side effects
Drugs that stop the breakdown of dopamine	Can be used with levodopa; usually given when dopamine replacement drugs start to lose their effectiveness
Anticholinergic drugs	Blocks the action of the ACh

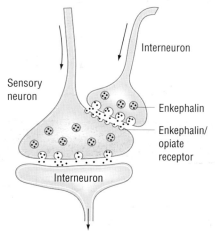

Drugs such as **diamorphine (heroin)** can be used to relieve severe pain. Diamorphine is a chemical derivative of morphine, but it is more soluble and penetrates the brain more rapidly.

✔*Quick check 3*

How might opioids block the transmission of pain?

Diamorphine hydrochloride belongs to a group of medicines called **opioids**. Opioids mimic the effects of naturally occurring pain-reducing chemicals – **endorphins**. They combine with the opioid receptors in the brain and block the transmission of pain signals, so less pain is actually felt.

✔ *Quick check 4*

Opioids attach to a specific membrane protein called an opioid receptor. Opioid receptors are located on sensory neurone cell membranes in large concentrations in the brain and spinal cord. A chain of reactions is stimulated which results in a depression of the normal activity of the neurone for a short time. They then leave the receptor and the normal function of the neurone returns.

Cannabis is derived from a plant. It is a central nervous system depressant. It is available in three main forms as:

- the dried leaves and buds, known as **marijuana** (or **grass**)
- a solid resin (**hashish** or **hash**) which is collected from the buds and flower heads
- a thick liquid or resin prepared from the flowers (**hash oil**); this is more powerful than the other forms of cannabis.

Uses of cannabis

- **Therapeutic use**: as a mild analgesic and sedative, which may relieve the symptoms of multiple sclerosis, hypoglycaemia and other disorders. It has been used as a medication for the terminally ill when other treatments have failed to relieve distress.

However, its possession or use in the UK is illegal at present and doctors are not able to prescribe cannabis in any form. It was legally prescribed until 1928.

- **Recreational use**: cannabis is the most widely used illegal drug in the UK. The effects of cannabis depend upon the amount used, its potency and the expectations of the user. The most common effects are talkativeness, cheerfulness, relaxation, and greater appreciation of sound and colour. It usually helps to relax the user. Perception of time, and occasionally of space, is altered.

✔ *Quick check 5*

Like other drugs cannabis has dangerous side effects, especially when higher doses are taken. These include hallucinations, disorientation, anxiety and depression.

Physical dependence results from repeated, heavy use of drugs like heroin and alcohol. This can change the body chemistry so that if someone does not get a repeat dose they suffer physical withdrawal symptoms.

Psychological dependence is more common and can happen with any drug. People who are psychologically dependent on heroin or alcohol find that using it becomes far more important than other activities in their lives. They crave the drug and will find it very difficult to stop using it, or even to cut down on the amount they use.

✔ *Quick check 6*

Dependence will often include both physical and psychological factors.

Hint
Opioid receptors are usually G proteins (guanine nucleotide binding proteins) The G protein moves within the membrane until it reaches its target. It may inhibit voltage-activated calcium channels so transmission of the pain impulse is prevented.

Hint
The plant *Cannabis sativa* is grown all round the world. The main mind-altering (psychoactive) ingredient in cannabis is THC (delta-9-tetrahydrocannabinol), but more than 400 other chemicals are present in the plant.

Hint
Cannabis is an illegal class B drug.

Hint
In 600 BC cannabis was mentioned as an intoxicating resin. It was believed to quicken the mind, prolong life, improve judgement, lower fevers, induce sleep and cure dysentery. Because of its psychoactive properties it was more highly valued than medicines with only physical activity.

QUICK CHECK QUESTIONS

1. What is a drug?
2. Which chemical needs to be replaced in people suffering with Parkinson's disease and why cannot it be taken directly by mouth?
3. Which drug can be used for the relief of severe pain?
4. How do opioids work?
5. Name one therapeutic and one recreational use of cannabis.
6. What is the main difference between physical and psychological dependency?

UNIT 5

The importance of homeostasis

Key words

- homeostasis
- set point
- receptor
- effector
- negative feedback
- autonomic nervous system
- hypothalamus
- core body temperature
- thermoreceptors

Examiner tip

Learn the definition of homeostasis!

Examiner tip

Make sure you do not confuse negative and positive feedback.

✓ *Quick check 3*

✓ *Quick check 4*

Key definition

Homeostasis is the maintenance of a stable internal environment.

Why is homeostasis important?

✓ *Quick check 1*

We must be able to adjust to changes in our environment because living cells can function only within a narrow range of:

- oxygen and nutrient availability
- temperature
- pH and ion concentration.

Conditions inside our bodies fluctuate within this narrow range about an average called a **set point**.

✓ *Quick check 2*

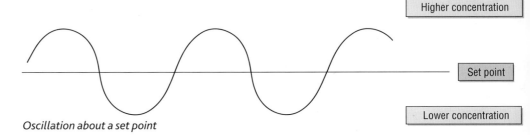

Higher concentration

Set point

Lower concentration

Oscillation about a set point

Homeostasis requires:

- a **receptor** that receives information about a change, e.g. a fall in blood pH
- an **effector** that responds and brings about changes in order to return to the set point, e.g muscles which increase ventilation rate to remove excess CO_2
- an efficient means of communication between the receptor and the effector
- continuous monitoring of the factor being controlled, e.g. chemoreceptors which monitor blood pH.

Negative feedback occurs when the body senses an internal change and then activates mechanisms that reverse the change.

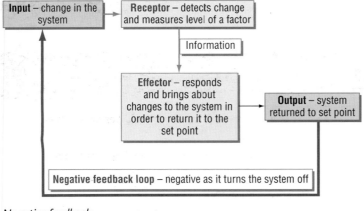

Negative feedback

Role of the autonomic nervous system in homeostasis

- The 'involuntary' or autonomic nervous system is responsible for homeostasis.
- These controls are done automatically below the conscious level.
- The autonomic nervous system consists of two sets of motor neurones carrying impulses to effectors.
- Neurones of the sympathetic and parasympathetic nervous systems have different, often opposing, effects. Example (see table over page):

- The **hypothalamus** has an important role in coordinating the autonomic nervous system.

	Sympathetic	Parasympathetic
Heart	Increased heart rate	Decreased heart rate
Bronchi	Dilates bronchi	Constricts bronchi
Liver	Increase in amount of glucose released	Causes the liver to increase production of glycogen

✓*Quick check 5*

Principles of homeostatic mechanisms

The flow diagram shows the negative feedback loops for temperature control in a mammal.

Our **core body temperature** usually remains within a narrow range, between 36.5 °C and 37.5 °C. This is important because this is the temperature at which the majority of the chemical reactions in the body work most efficiently. Too high and enzymes can denature, too low and reaction rates are too slow.

The temperature will fluctuate, about a set point, depending on activity and time of day.

A change in the environmental temperature is detected by **thermoreceptors** in the skin. The thermoregulation centre in the hypothalamus receives information from the thermoreceptors in the skin and internal organs, as well as detecting changes in the temperature of blood flowing through the brain. If a change is detected, the hypothalamus sends impulses to the various effectors which will reverse the changes.

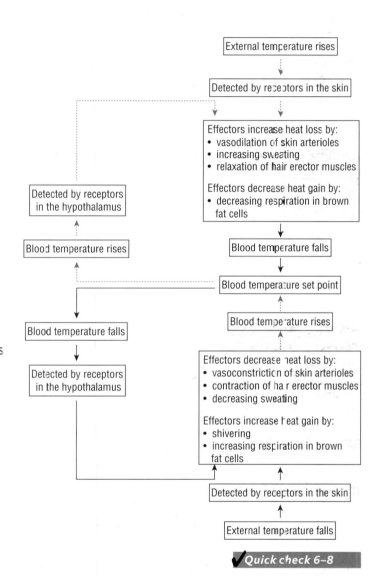

✓*Quick check 6–8*

QUICK CHECK QUESTIONS

1 What do you understand by the term 'homeostasis'?

2 Why do organisms need a communication system to respond to changes in the environment?

3 What is the difference between a receptor and an effector?

4 Explain what is meant by negative feedback in a living organism.

5 What is the role of the autonomic nervous system in homeostasis?

The next questions refer to the flow diagram above.

6 What is the set point in the feedback loop for temperature control?

7 How do effectors reduce heat loss in a human?

8 Where are the receptors located that detect a change in blood temperature?

Control of temperature

Key words

- vasodilation
- vasoconstriction
- thyroxine
- metabolic rate
- core body temperature
- hypothalamus
- hypothermia
- hyperthermia

✔ Quick check 1 and 2

Structure	Body too hot	Body too cold
Arterioles in the skin	dilate so more blood flows through them and heat is lost by radiation and convection – **vasodilation**	constrict so less heat lost through radiation – **vasoconstriction**
Sweat glands	secrete sweat which evaporates using the heat energy in the skin – cools the body	No sweat secreted
Erector muscles	Do not contract	Contract, pulling the hairs vertically up forming a pocket of air under each hair. Air is a bad conductor of heat so insulates the body, reducing heat loss
Muscles		Contract and relax – body shivers increasing respiration in muscle and releasing more heat energy

Examiner tip

Think about why increased respiration should result in an increase in temperature.

✔ Quick check 3

Long-term control of temperature

Thyroxine increases the **metabolic rate**, which increases the release of heat energy. Thyroxine works by stimulating the transcription of genes which control the synthesis of more respiratory enzymes. If we are exposed to cold conditions for more than a few days we secrete extra thyroxine – we become acclimatised.

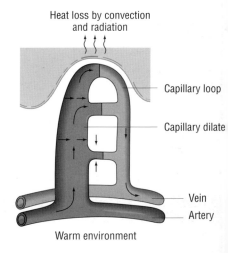

How and why is core body temperature measured?

It is important to measure **core body temperature**:

- because a number of diseases are accompanied by characteristic changes in body temperature, e.g. fevers
- as a way of monitoring the course of certain diseases
- as a means for the doctor to evaluate the efficiency of a treatment.

✔ Quick check 4

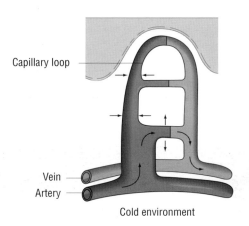

Vasodilation and vasoconstriction

Temperature is measured using a thermometer.

- **Electronic thermometers** – work by measuring how well electricity travels through a wire. Electronic thermometers are used in the mouth, rectum or armpit. They are accurate and are easy to use and read.
- **Ear thermometers** – use infrared energy to measure body temperature which is shown on a digital display. The measurement of ear temperature accurately reflects core body temperature because the eardrum shares its blood supply with the temperature control centre in the hypothalamus.
- **Forehead thermometers** use skin temperature to determine body temperature. Press the strip against a person's forehead – the temperature will make some of the numbers change colour. Not as accurate as electronic and ear thermometers.

> **Hint**
> Accuracy is an assessment of how close an observed value is to the true value.

Hypothermia

Hypothermia occurs when the body's core temperature drops below 35 °C. As the core temperature falls all metabolic reactions slow down. This is because molecules have less kinetic energy so there will be fewer collisions between enzymes and their substrates.

Symptoms	Who is at risk?	Treatment
Mild hypothermia: shivering and muddled thinking **Moderate hypothermia**: violent shivering, inability to think, weak pulse, slow and shallow breathing **Severe hypothermia**: shivering stops, difficulty in speaking, little or no breathing	Those who spend a lot of time outdoors in cold weather The elderly, as may not be very mobile and cost of heating is high Babies aged under 1 year Those who are intoxicated	Call for emergency medical help Monitor breathing: expired air resuscitation if needed Cardiopulmonary resuscitation (CPR) if pulse/heart stops Move person out of the cold, or cover head and insulate body from cold ground Replace wet clothing with dry blankets

Hyperthermia

✓ *Quick check 5*

✓ *Quick check 6*

Hyperthermia occurs when the core body temperature rises significantly above normal.

Symptoms	Who is at risk?	Treatment
Mental confusion, headache, muscle cramps and vomiting. May act drunk or combative	Infants and young children – do not leave them in a parked car; dress them in loose clothing and a hat People over age of 65 Overweight people People who over-exert during work or exercise	Advanced state = heat stroke which requires hospitalisation Goal is to lower body temperature quickly – take the person to a cool place, bathe in cool water, cover with cold towels, use a fan to wick heat away from the skin. Patient must also be hydrated – intravenous drip/isotonic drinks

✓ *Quick check 7*

QUICK CHECK QUESTIONS

1. How does vasodilation cool the skin?
2. Explain how vasoconstriction helps to reduce heat loss.
3. How do we become acclimatised to living in cold conditions?
4. What is the importance of measuring core body temperature?

5. How would you recognise and treat hypothermia?
6. What is the difference between hypothermia and hyperthermia?
7. How would you recognise and treat hyperthermia?

Diabetes

Key words

- islets of Langerhans
- pancreas
- endocrine gland
- glucagon
- insulin
- glycogen
- glycogenesis
- gluconeogenesis
- type 1 diabetes
- type 2 diabetes

Key definition

Endocrine glands are glands that secrete hormones directly into the blood rather than through a duct.

Some cell clusters, known as **islets of Langerhans**, in the **pancreas**, secrete hormones which regulate the level of glucose in the blood. ✔ *Quick check 1*

The islets contain two types of endocrine cell scattered throughout the exocrine tissue:

- alpha cells (α cells) – secrete **glucagon**
- beta cells (β cells) – secrete **insulin**.

✔ *Quick check 2 and 3*

Alpha cells are usually distributed around the edge of the islet and beta cells in the centre.

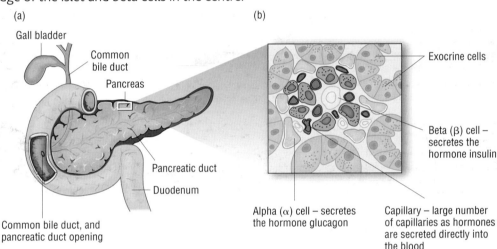

Exocrine cells

Islet of Langerhans

Endocrine ce

(a)

Gall bladder
Common bile duct
Pancreas
Pancreatic duct
Duodenum
Common bile duct, and pancreatic duct opening

(b)

Exocrine cells

Beta (β) cell – secretes the hormone insulin

Alpha (α) cell – secretes the hormone glucagon

Capillary – large number of capillaries as hormones are secreted directly into the blood

a The pancreas; b islets of Langerhans

Negative feedback controls blood glucose concentration

If blood glucose levels **rise**, e.g. following a meal, then:

- both α and β cells detect the rising concentration of blood glucose (**receptors**)
- α cells stop secreting glucagon but β cells secrete insulin into blood plasma
- insulin binds to receptors on cell surface membrane of liver, fat and muscle cells
- uptake of glucose by these cells is increased (**effectors**)
- use of glucose in respiration is increased
- liver cells convert glucose to **glycogen** (**glycogenesis**) which is stored
- blood glucose levels fall.

If blood glucose levels **fall** due to glucose being used rapidly by the cells or lacking in the food eaten then:

- α and β cells detect the falling concentration of blood glucose (receptors)
- β cells stop secreting insulin
- α cells start to secrete glucagon into the blood plasma
- less glucose is taken up by the target cells (**effectors**)
- rate of the use of glucose decreases
- **gluconeogenesis** occurs – amino acids, pyruvate and lactate are converted to glucose for respiration
- glycogen is converted to glucose in liver cells and released into the blood
- blood glucose levels rise.

✔ *Quick check 4 and 5*

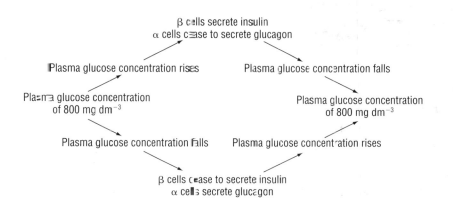

β cells secrete insulin
α cells cease to secrete glucagon

Plasma glucose concentration rises → Plasma glucose concentration falls

Plasma glucose concentration of 800 mg dm^{-3} → Plasma glucose concentration of 800 mg dm^{-3}

Plasma glucose concentration falls ← Plasma glucose concentration rises

β cells cease to secrete insulin
α cells secrete glucagon

Types of diabetes

Diabetes usually refers to **diabetes mellitus**. There are two forms.

Type 1 (insulin dependent)	Type 2 (insulin independent)
Body is unable to produce insulin because the β cells are destroyed by its own immune system	The body target cells lose their responsiveness to insulin
Normally begins in childhood (juvenile onset diabetes)	Usually starts when over 40 years old (late onset diabetes)
Risk factors: autoimmune disease; the immune system possibly triggered by a virus; an allele has been discovered that increases a person's risk of developing type 1 diabetes	**Risk factors:** age; obesity; family history; sedentary lifestyle; high blood pressure; low birth weight – lack of nutrients in the womb destroys insulin-producing cells in the pancreas
Treatment: insulin injections using a **syringe** twice or four times daily; an **insulin pen** – needle in the tip penetrates just under the skin delivering the required amount of insulin; an **insulin pump** for continuous insulin delivery, 24 h a day	**Treatment:** can usually be controlled by diet; oral drug treatment preferably by metformin, otherwise by sulphonylureas with insulin – used to stimulate insulin production

Role of the diabetes nurse

The diabetes nurse provides education, advice and support to diabetics. This includes:
- crisis management advice
- support and advice during periods of change of treatment
- assistance with day-to-day management of patients admitted to hospital.

✔ *Quick check 6*

Why are regular checks important?

Blood pressure – major cause of death for diabetics is coronary heart disease.

Examination of the retina – poorly controlled diabetes damages the tiny blood vessels in the retina. This may cause blurring and occasionally loss of vision.

Kidney function tests – a diabetic may hold excess sugar in their urine so more likely to develop bladder infections such as cystitis. Diabetes can also affect the glomeruli (capillaries).

✔ *Quick check 7*

QUICK CHECK QUESTIONS

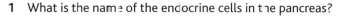

1 What is the name of the endocrine cells in the pancreas?
2 What do the alpha cells secrete?
3 What do the beta cells secret?
4 Describe what happens if blood glucose is too high.

5 Describe how blood glucose concentration is raised.
6 Distinguish between type 1 and type 2 diabetes mellitus.
7 Why are regular checks so important for a person with diabetes mellitus?

UNIT 5

Excretion of nitrogenous waste

Key words

- excretion
- metabolism
- deamination
- nitrogenous waste
- cortex
- medulla
- pelvis
- nephron
- Bowman's capsule
- glomerulus
- proximal and distal convoluted tubules
- loop of Henlé
- collecting duct

Examiner tip

Make sure that you do not confuse excretion with elimination – elimination (egestion) is the removal of waste products, such as fibre, that have never been involved in metabolism.

✓*Quick check 1*

Excretion is the removal of waste products of **metabolism** from the body.

Key definition

Metabolism: the chemical processes occurring within a living cell or organism that are necessary for the maintenance of life.

Why is excretion important?

It removes toxic (poisonous) waste from the body. If these toxins build up they will slow down and eventually stop important chemical reactions taking place.

One of the waste products of respiration is **carbon dioxide**:

   sugar + oxygen → energy + water + carbon dioxide

Carbon dioxide reacts with water in the blood to form carbonic acid which dissociates into hydrogen ions and hydrogen carbonate. Accumulation would cause the acidity of the blood to increase – particularly harmful to the cells of the brain. Eventually the person would become unconscious and may die. If the carbon dioxide in the lungs increases by only 0.2% from a normal level then breathing rate is doubled.

We cannot store proteins or amino acids in our bodies, so any excess is broken down into fats or carbohydrates. It is the amino group that has to be removed by a process called **deamination** in the liver (removal of **nitrogenous waste**).

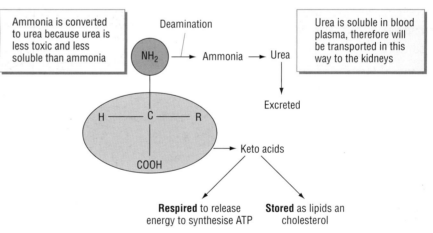

Diagram to summarise breakdown of an amino acid

Kidney structure

The urinary system is made up of:

- kidneys – one quarter of our blood supply passes through the kidneys every minute
- ureters – carry urine from the kidney to the bladder
- bladder – stores urine
- urethra – takes the urine from the bladder to the outside.

Each kidney receives blood through the renal artery, which branches off the aorta (main artery). When the blood has been filtered by the kidneys it returns to the bloodstream through the renal vein, which connects to the vena cava.

✓*Quick check 2*

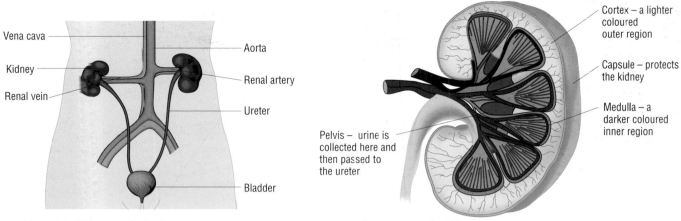

Vena cava

Kidney

Renal vein

Aorta

Renal artery

Ureter

Bladder

Position of the kidneys in the body

Cortex – a lighter coloured outer region

Capsule – protects the kidney

Medulla – a darker coloured inner region

Pelvis – urine is collected here and then passed to the ureter

Longitudinal-section of a kidney

There are three regions inside a kidney:

- **cortex** – outer region
- **medulla** – the inner part of the kidney
- **pelvis** – the funnel-shaped cavity that receives the urine. ✔ *Quick check 3*

- Microscopic examination shows that the cortex and medulla are made up of around one million tiny tubules or **nephrons** and their associated blood vessels. The nephron is basically a hollow tube with an opening at each end.

- **Bowman's capsule (renal capsule)** is at one end of the tubule – a double-walled cup-like structure enclosing a cluster of capillaries (**glomerulus**).

- The highly coiled **proximal convoluted tubule** is next (in the cortex).

- The **loop of Henlé** follows – a U-shaped tube with an ascending and a descending limb.

- Last is the **distal convoluted tubule** which leads to the **collecting duct**.

✔ *Quick check 4, 5 and 6*

> **Hint**
>
> **Histology of the kidney**: you need to be able to identify the structures in the nephron in photomicrographs of kidney sections.

Renal capsule

Proximal convoluted tubule

Distal convoluted tubule

Efferent arteriole

Afferent arteriole

Glomerulus

From renal artery

Descending limb of loop of Henle

Ascending limb of loop of Henlé

Cortex

Medulla

Collecting duct

Pelvis

Nephron

QUICK CHECK QUESTIONS

1 Why is excretion essential for humans?

2 What is the name of the vessel through which the kidney receives blood?

3 There are three regions in a kidney. What are they?

4 State the parts of a nephron that can be found in a kidney cortex.

5 What is the name of the cluster of capillaries found inside a Bowman's capsule?

6 How would you describe the loop of Henlé?

How a kidney works

There are two stages:

- **Ultrafiltration** – filtering of small molecules out of the blood and into the **renal capsule**.
- **Reabsorption** – taking back any useful molecules from the fluid in the nephron.

Ultrafiltration

- The **afferent arteriole** supplying the glomerulus has a larger diameter than the **efferent arteriole** carrying blood away so pressure is built up. This high **hydrostatic pressure** (filtration pressure) forces blood against a filter.
- The endothelium of the blood capillaries has gaps between the cells through which plasma can escape.
- The **basement membrane** (filtration membrane) acts as a molecular filter. No proteins with a molecular mass greater than 69 000 and no cells can get through (so they stay in the blood).
- The inner layer of the renal capsule consists of **podocytes** – cells lifted off the surface on little 'feet' so the filtrate passes beneath them and through gaps between their branches into the capsule.

Wide afferent arteriole

Narrow efferent arteriole

Capillaries of glomerulus

Cavity of renal capsule

Nucleus of podocyte

Filtration slit

Podocyte in wall of renal capsule

Basement membrane

Endothelial cell

Capillary lumen

Circular pore

→ Filtrate is forced out of capillary by the hydrostatic pressure of the blood in the glomerulus. This pressure is sufficient to overcome the pressure of the fluid in the renal capsule

Ultrafiltration in the renal capsule

✓ Quick check 1

Selective reabsorption

	Proximal convoluted tubule	Loop of Henlé	Distal convoluted tubule
What is reabsorbed into the blood?	About 85% of the filtrate is reabsorbed here – all the glucose, amino acids, vitamins and hormones, some of the sodium and chloride ions, and some water	Water by osmosis (passive)	Sodium ions – by active process
Special adaptation	Cells covered in **microvilli** Large number of mitochondria. Co-transporter proteins in the cell surface membrane – facilitated diffusion	See diagram overpage	Potassium ions are actively transported in the opposite direction. Rate of transportation of both ions can be varied to maintain a homeostatic balance

✓ Quick check 2, 3, 4 and 5

Examiner tip

Check that you are sure about the difference between active and passive transport.

Control of water balance

✓ Quick check 6

A **negative feedback loop** consisting of a **receptor** (receptor cells in the **hypothalamus**) that monitors the amount of water in blood plasma, and an **effector** (the **pituitary gland**, producing ADH if water is too low, and the **walls of the distal convoluted tubule**) that returns the water content of the blood to normal.

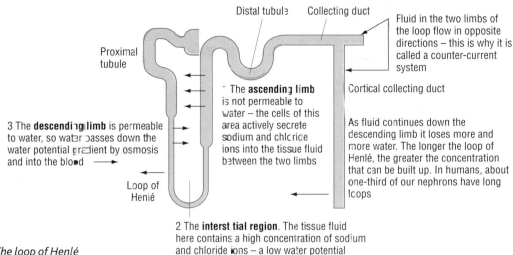

The following labels appear on the diagram:

Distal tubule   Collecting duct

Fluid in the two limbs of the loop flow in opposite directions – this is why it is called a counter-current system

Proximal tubule

The **ascending limb** is not permeable to water – the cells of this area actively secrete sodium and chloride ions into the tissue fluid between the two limbs

Cortical collecting duct

3 The **descending limb** is permeable to water, so water passes down the water potential gradient by osmosis and into the blood →

As fluid continues down the descending limb it loses more and more water. The longer the loop of Henlé, the greater the concentration that can be built up. In humans, about one-third of our nephrons have long loops

Loop of Henlé

2 The **interstitial region**. The tissue fluid here contains a high concentration of sodium and chloride ions – a low water potential

The loop of Henlé

Regulation of water potential of the blood

- The water potential of our blood is monitored by **osmoreceptors** (cells) in the
- hypothalamus of the brain.
- If more water is lost from the body than gained, a fall in water potential of blood flowing through the hypothalamus is detected. The result is a loss of water from cells by osmosis; the osmoreceptor cells shrink.
- This change stimulates neurosensory cells in the hypothalamus to produce and secrete **antidiuretic hormone** (ADH).
- ADH passes along the axons of the neurosensory cells which end in the **posterior pituitary gland** from where ADH is released into the blood plasma.

How does ADH affect the kidneys?

- ADH travels in the blood to the kidneys where it binds to receptors on the cell surface membrane. This increases the permeability to water of the cell surface membrane of the endothelial cells in the walls of the distal convoluted tubule and the collecting duct.
- Receptors on the cell surface membrane pick up the ADH molecules causing an increase in **cyclic AMP** (cAMP).
- cAMP leads to the insertion of **aquaporins** into the collecting duct cell membrane, where they open, making the membrane more permeable to water.
- Water leaves the collecting duct by osmosis and is reabsorbed into blood plasma.
- The fluid in the collecting duct has become more concentrated so a smaller quantity of more concentrated urine is produced.
- The osmoreceptors in the hypothalamus detect the rise in water potential, and stop sending impulses to the pituitary gland, which reduces the release of ADH.

Key definition

Aquaporins are intrinsic proteins that form pores through the phospholipid bilayer. They selectively conduct water molecules in and out, while preventing the passage of ions and other solutes. Also known as water channels.

✔ *Quick check 7*

QUICK CHECK QUESTIONS

1 What is the role of the basement membrane in ultrafiltration?
2 What is the importance of the microvilli in the proximal convoluted tubule?
3 Why do you think that there are many mitochondria present in the cells of the proximal convoluted tubule?

4 Explain why the process of selective reabsorption is necessary in the kidney.
5 What is the difference between active and passive reabsorption? Give an example of each.
6 State the roles of the hypothalamus and the posterior pituitary gland in osmoregulation.
7 What is the role of cyclic AMP?

UNIT 5

More on the kidney

Key words

- diabetes insipidus
- diabetes mellitus
- renin
- angiotensin
- erythropoietin
- creatinine
- glomerular filtration rate
- oedema
- hypertension

✔ Quick check 1

What do changes in the chemistry of the urine indicate?

- Dark urine is a sign of dehydration whereas a large volume of clear urine indicates over-hydration, which is usually considered much healthier than dehydration.
- If a person excretes very large volumes of dilute urine, even when the blood water level is low, it could indicate **diabetes insipidus**.
- Normally there is no glucose in the urine because it is reabsorbed by active uptake in the nephron. Glucose only enters the urine when blood glucose level reaches above 180 mg/dl, indicating that the patient is suffering from **diabetes mellitus**.

Differences between diabetes mellitus and diabetes insipidus

	Diabetes mellitus ('sugar' diabetes)	Diabetes insipidus ('water' diabetes)
What it is	Inability to control blood glucose levels	Cannot control the balance of water in the body
Test	Use a Test Strip, e.g. Clinistix, to test glucose levels in urine	Specific gravity checks the concentration of substances in the urine
Causes	Type 1 diabetes: the pancreas does not produce sufficient insulin	Damage to hypothalamus so malfunction in antidiuretic hormone (ADH) secretions
	Type 2 diabetes: the pancreas does not secrete enough insulin and the body cannot use it properly	Genetic disease – a mutation in a gene coding for the production of aquaporins or for the receptors on the collecting duct membrane
Treatment	Type 1 diabetics use insulin injections	Give the patient synthetic ADH
	Type 2 diabetics use careful control of diet and possibly insulin injections	Give the patient a low sodium diet and medication to reduce the amount of urine produced

✔ Quick check 2

Other homeostatic mechanisms in the kidney

The kidney is one of the major organs involved in homeostasis.

Hint

What is the difference between hypertension and hypotension?

✔ Quick check 3

Renin (angiotensinogenase): When blood pressure is low, the kidneys secrete renin. Renin stimulates the production of **angiotensin** which causes blood vessels to constrict, resulting in increased blood pressure. Damaged kidneys may release too much renin. High blood pressure increases the risk of heart attack, congestive heart failure and stroke.

Key definition

Anaemia is when your body does not have enough red blood cells.

Erythropoietin: When oxygen levels are low (e.g. at high altitude), specialised cells in the kidney secrete erythropoietin. A diseased kidney produces less erythropoietin so the patient could become anaemic. Erythropoietin stimulates the bone marrow to produce more erythrocytes and so increase the oxygen-carrying capacity of the blood.

✔ Quick check 4

Kidney failure

When the kidneys start to fail, toxins, including urea and other nitrogenous waste, are not filtered out of the blood and start to accumulate in the tissues.

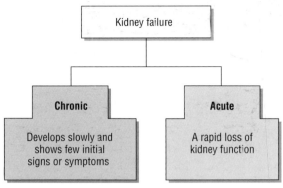

Two main types of kidney failure

Causes of kidney failure

Acute renal failure	Chronic renal failure
• serious illness or operation • sudden loss of large amounts of blood or tissue fluid • sudden blockage for example, a kidney stone • side effect of some medications • bacterial infection in the kidneys	• complication of diabetes mellitus • uncontrolled high blood pressure • an inflammation affecting the kidney tissue • the result of certain inherited diseases, such as polycystic kidney disease

Diagnosing kidney failure

Urine samples may show the presence of red blood cells, white blood cells and proteins.

Blood samples may show high levels of **creatinine**, which increase as kidney disease progresses. **Glomerular filtration rate** (GFR) is a measure of how well the kidneys are removing wastes and excess fluid from the blood.

The most **characteristic symptoms** of kidney failure are a reduction in the volume of urine and **oedema**.

Ultrasound scans and CT scans of the kidneys will identify any structural abnormalities or blockage (such as kidney stones) of the kidneys.

Kidney and cardiovascular disease go hand-in-hand

- Cardiovascular disease is common in people with chronic kidney disease. This can be due to anaemia, **hypertension**, diabetes.
- The increase in blood pressure caused by the increased production of renin is a risk factor in cardiovascular disease. Renin acts by causing vasoconstriction leading to hypertension.

✔ *Quick check 5*

Key definition

Oedema (fluid retention): feet, hands and the area around the eyes may swell and become puffy due to fluid retention in tissues.

✔ *Quick check 6*

Key definition

The haematocrit measures how much space in the blood is occupied by red blood cells. It measures the ratio of the volume occupied by packed red blood cells to the volume of the whole blood.

Hint

Check the use of the haemocytometer. Think about the advantages and disadvantages of each technique.

QUICK CHECK QUESTIONS

1 What can changes in the chemistry of urine indicate?

2 Distinguish between diabetes mellitus and diabetes insipidus.

3 What can overproduction of renin cause? What is this an example of?

4 When is erythropoietin secreted by the kidney?

5 State two causes of acute and two of chronic renal failure.

6 How would the presence of a kidney stone be diagnosed?

Treatment of kidney disease

Treatments available

- erythropoietin drugs to curb **anaemia**; drugs to help control calcium and phosphorus levels
- special diet to decrease workload on the kidneys, keep body fluids and minerals in balance, and stop a build-up of toxins, e.g. low protein diet to reduce urea
- blood pressure control – important in order to prevent or delay the progression of kidney disease
- renal dialysis
- transplant

✔ *Quick check 1*

Renal dialysis

This is an artificial method of maintaining the chemical balance of the blood when the kidneys have failed.

1 **Haemodialysis** – blood is circulated through a machine, containing a membrane, and back into the patient via catheters. Haemodialysis can be carried out in a special unit in a hospital or at home.

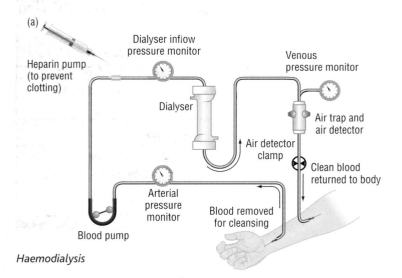

(a)

Heparin pump (to prevent clotting)

Dialyser inflow pressure monitor

Venous pressure monitor

Dialyser

Air trap and air detector

Air detector clamp

Clean blood returned to body

Arterial pressure monitor

Blood removed for cleansing

Blood pump

Haemodialysis

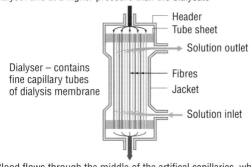

(b) Blood inlet via a pump, which keeps the blood moving through the dialyser and at a higher pressure than the dialysate

Header
Tube sheet
Solution outlet

Dialyser – contains fine capillary tubes of dialysis membrane

Fibres
Jacket

Solution inlet

Blood flows through the middle of the artifical capillaries, while dialysing fluid flows along the outside in the opposite direction

2 **Peritoneal dialysis** – dialysing fluid is circulated into and out of the abdomen using the patient's peritoneal membrane as a filter.

✔ *Quick check 4*

Peritoneal dialysis works on the principle that the peritoneal membrane surrounding the intestine can act as a partially permeable membrane. Dialysis solution is poured in and fills the space between the abdomen wall and the organs. Dialysis occurs by diffusion. Excess water is removed by osmosis by altering the concentration of glucose and solutes in the fluid. Peritoneal dialysis can be done in the patient's home, workplace or almost anywhere.

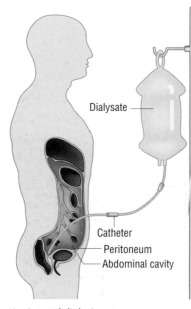

Dialysate

Catheter
Peritoneum
Abdominal cavity

Peritoneal dialysis

Use of recombinant erythropoietin in kidney failure and dialysis

In kidney failure, the production of **erythropoietin** is affected and new red blood cells are made less quickly. This leads to anaemia and blood loss – as residual blood in the dialyser or blood lines, through frequent blood testing or through a dialyser catheter when blood is discarded routinely at the start of dialysis. These problems can be overcome by giving the patient recombinant erythropoietin.

✔ *Quick check 6*

Kidney transplants

Renal dialysis helps:
- to replace the work that the kidneys did
- the patient to feel better and live longer

but does not cure kidney failure.

Kidney transplantation surgically places a healthy kidney from another person into the body. The transplanted kidney does the work that the two failed kidneys used to do.

It is essential that the tissue type of the donor matches that of the recipient or the kidney will be rejected by the recipient's immune system.

Examiner tip

Check your knowledge of the immune reaction.

What are the advantages of transplantation over renal dialysis?

Transplantation	Renal dialysis
There is no dependence on a machine for the rest of one's life	No use of rejection medicines
Increased energy; usually anaemia (feeling of tiredness) is reversed after a successful transplant	For patients on dialysis, correction of anaemia requires life-long use of erythropoietin injections, which are extremely expensive
Diet is less limited; fluid intake is less restricted	Diet is limited; control amount of salt (sodium chloride), potassium and possibly fluid intake
A better quality of life; can go back to work in a style similar to that before the illness	Have to undergo dialysis several times a week
A woman may be able to conceive a child after having a successful transplant	Women on dialysis usually do not ovulate and therefore are unable to become pregnant

✔ *Quick check 7*

Practical and ethical issues involved in use of donor organs

- Patients wait approximately 3 years for a kidney transplant.
- Demand for kidney transplantations is increasing because the number of people with severe kidney disease is increasing.
- Only 1 in 4 sign the donor register.
- 'Opting out not opting in' suggested but this would remove individual's fundamental right to decide. Most European countries use the system of 'presumed consent'.
- Doctors have to ask a family who have just watched someone die for the body parts.
- Solve supply problem by buying from a live donor?

QUICK CHECK QUESTIONS

1 What treatments are available if your kidneys are diseased?
2 In haemodialysis, blood and dialysate flow in opposite directions. Why?
3 Why is heparin added to the blood in the dialyser?
4 What is the difference between haemodialysis and peritoneal dialysis?

5 Why is a partially permeable membrane essential in dialysis?
6 Why is it important that patients on dialysis are given recombinant erythropoietin?
7 What is the main advantage of a person with kidney failure receiving a transplant?

Effects of ageing on the reproductive system

Key words

- oocytes
- FSH
- menopause
- testosterone
- oestrogen
- progesterone
- cholesterol
- CHD
- HRT
- progestin
- hysterectomy
- phytoestrogens
- antioxidants

✓ *Quick check 1*

With age our bodies deteriorate.

In women:

- the chance of becoming pregnant decreases with age
- the risk of miscarriage increases with age
- the quality of **oocytes** declines with age
- follicle-stimulating hormone (**FSH**) levels increase with age
- **menopause** occurs (see below).

In men:

- there is a reduction in the number and motility of sperm produced
- there is a reduction in the percentage of normal sperm produced
- levels of **testosterone** reduce
- there is a decrease in sex drive
- frequency, length and rigidity of erections decline
- blood flow in penis is reduced together with the volume of the ejaculate.

In both:

- there is reduced reproductive capacity.

The menopause

The average age for menopause to occur is between 45 and 55 years. A woman's periods stop and she is no longer able to conceive.

Changes during menopause

- At first (the peri-menopausal period), the regular cycle of menstruation becomes disrupted; periods become irregular. When the periods stop completely this is known as the menopause.
- **Oestrogen** production slows and eventually stops.
- Less **progesterone** is made.
- Transient (short-term) changes include hot flushes, fatigue, anxiety, memory loss and sleep disturbance.

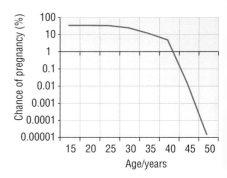

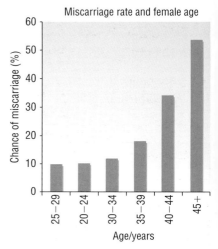

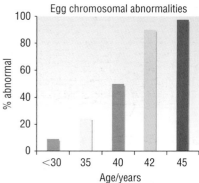

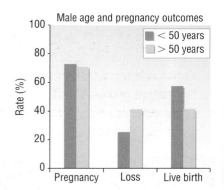

The effect of ageing on the reproductive process

Changes in a woman's physiology due to lack of oestrogen and progesterone

- Levels of **cholesterol** increase. This increases the risk of coronary heart disease (**CHD**).
- For the first 5 years after the start of menopause calcium loss from the bones increases resulting in a loss of bone density.
- The percentage of body fat increases and muscle mass decreases leading to a change in body composition.

Examiner tip

Check that you still understand the link between cholesterol and CHD.

✓*Quick check 2*

Use of HRT in treating the symptoms of menopause

Hormone replacement therapy (**HRT**) reverses the decline in levels of oestrogen and so can relieve symptoms such as hot flushes and sleep disturbances.

Types of HRT

Type	Who should use it	Which hormones this type of HRT contains
Oestrogen only	Women who have had their uterus removed (**hysterectomy**)	Oestrogen
Cyclic (continuous sequential)	Women who have not had a hysterectomy and who have not stopped having periods	Tablets contain both oestrogen and progesterone two different coloured tablets in the pack. One tablet contains oestrogen on its own. The other contains both oestrogen and progestin, so they receive progesterone for only part of the cycle
Continuous combined	Women who have not had a hysterectomy and are post-menopausal	Oestrogen and progesterone are both taken daily

Advantages and disadvantages of different forms of HRT

HRT	Advantages	Disadvantages
Tablet	Easy to take Easy to top Easy to adjust dose	Need to remember to take every day May interfere with other medication
Patches	Low dose hormones go straight into bloodstream No need to take tablets	May cause skin irritation May be visible May fall off

Key definition

Progestin is synthetic progesterone.

Side effects

✓*Quick check 3 and 4*

For short-term treatment the benefits of HRT in treating menopausal symptoms can outweigh the problems. If used for a longer time the risk of breast cancer, endometrial cancer, blood clots and stroke increases.

Alternative methods of treatment

- **Phytoestrogens** – found in plants; share some of the same biological activities with oestrogens produced in the body.
- **Antioxidants** – may protect your cells against the effects of free radicals. Found in fresh fruit and vegetables.

Examiner tip

Exam marks will be lost if you confuse the functions of the reproductive hormones (FSH, oestrogen, progesterone, testosterone).

✓*Quick check 5*

QUICK CHECK QUESTIONS

1. State six effects of ageing on the reproductive system.
2. What effect does the menopause have on a woman's physiology?
3. Why is HRT prescribed for some women during the menopause?
4. Why do you think some women prefer to wear a patch rather than take tablets?
5. What are the side effects of taking HRT?

Effects of ageing on the nervous system

Effects of ageing on the brain

- There is a decrease in brain weight and volume.
- The white matter decreases.
- The ventricular system enlarges.
- The brain generates fewer **neurotransmitters**.

Mental capacity does decrease with age but there is great individual variation. Possibly the loss of mental function is due to a combination of factors such as age, lack of mental stimulation, lack of social interactions, unhealthy diet, etc.

Alzheimer's disease is a degenerative disease of the nervous system. There are **three overlapping** stages.

✓ *Quick check 1*

First stage: memory loss becomes noticeable.

Second stage: mental ability declines, personality changes, cannot communicate coherently.

Third stage: needs constant supervision, depends totally on others, eventually language and ability to swallow are lost.

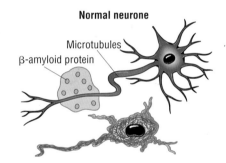

Normal neurone

Microtubules

β-amyloid protein

Neurone with neurofibrillary tangles

Neurofibrillary tangles

Pathology
When tissues from the brains of Alzheimer's patients are viewed under a microscope the major features are neurofibrillary tangles (composed mainly of Tan protein) and senile plaques.

Causes – many factors contribute to its onset:

- age – most important known risk factor
- family history
- severe head or whiplash injuries
- smoking, high blood pressure, high cholesterol
- gene responsible for production of **beta-amyloid protein** (link with Down's syndrome?).

Care of patients with dementia

This entails an emotional, physical, social and financial burden. The patient eventually may not recognise their carer.

✓ *Quick check 2*

Effect of ageing on the peripheral nervous system

Most noticeable changes with age are in our **senses:**

Hearing

Ongoing loss of hearing is linked to changes in the inner ear. Speech becomes difficult to understand, especially when there is background noise.

✓*Quick check 3*

Sight

Lens loses its elasticity as proteins in the cells begin to denature. This makes **accommodation** difficult.

Eye problem	What has happened	Symptoms	Causes	Treatment
Cataract	Clouding of the normally clear lens	Blurring of vision, sensitivity to light, poor night vision, difficulty with distinguishing colour	Age is the main cause – 99% of people in their 90s have cataracts Smoking Diabetes	Surgically remove the cloudy lens and replace it with an artificial lens
Glaucoma	Flow of fluid out of the eye is obstructed – causes a build up of pressure in the eye which damages the optic nerve and the nerve fibres	There is no pain and eyesight will seem to be unchanged and the eye may appear normal, but vision is being damaged	Age – affects 5% of people over 65 Race – Africans are more at risk Genetics	There is no cure but can usually be controlled by eye drops or tablets, so preventing or slowing further loss of sight. People over 40 are advised to have their eyes tested every 2 years to check for glaucoma
Age-related **macular degeneration** (AMD)	Two types – wet and dry Wet AMD is due to new blood vessels growing behind the retina, causing bleeding and scarring – leads to sight loss Dry AMD is more common – results from the ageing and thinning of macular (central) tissues of the retina	Loss of central vision. It affects the ability to read and to see fine detail	Exact cause of AMD is not known. Risk factors identified include: age, smoking, excess sunlight, high blood pressure, genetics	Treatment involves glasses; using bright lights, e.g. halogen lights; use of large-print books and a magnifying lens Some wet AMD patients may be treated by injecting a drug such as Lucentis into the back of the eye or by laser treatment

✓*Quick check 4, 5 and 6*

QUICK CHECK QUESTIONS

1 What is Alzheimer's disease and what are the possible causes of this disease?

2 What are the main issues involved in the care of patients with dementia?

3 What is the main effect of age on our sense of hearing?

4 What are cataracts and how is a cataract treated?

5 Why is it very important to have regular eye tests as we get older?

6 Why do people suffering with macular degeneration need large-print books and good lighting?

Effects of ageing on skeletal, cardiovascular and respiratory systems

Effects of ageing on the skeletal system

- Bone mass starts to reduce between the age of 30 and 40 years. **Osteoblast** activity declines but **osteoclast** activity continues at normal level.
- Bones break more easily and do not repair well.
- Vertebrae may collapse leading to 'Dowager's hump'.
- **Intervertebral discs** shrink – compressed discs and loss of bone mass lead to decrease in body height.
- Incidence of **osteoarthritis** increases.
- **Osteoporosis** – bones affected are less dense than normal bone so fractures are more likely.

✓ *Quick check 1, 2 and 3*

How can osteoporosis be prevented?

It is important to build up bone mass from childhood onwards.

- Drink milk or eat milk-based products so have sufficient calcium and vitamin D.
- Exercise is good, particularly weight-bearing exercise.
- Do not smoke and avoid excessive intake of alcohol.
- Check home safety to prevent falls.
- Post-menopausal women may need to replace oestrogen by taking HRT.
- Long-term use of some medications can lead to a loss of bone density.

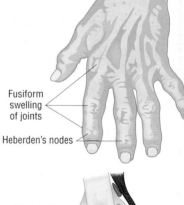

Fusiform swelling of joints

Heberden's nodes

How can osteoporosis be detected?

- 'DEXA' scan – the best method of detection. DEXA scans use X-rays to determine the density of bone.
- Ultrasound of the heel bone is also used to determine bone density. It uses cheaper equipment, but it is not yet clear if it is as accurate or reliable as DEXA scanning.

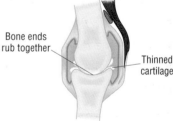

Bone ends rub together

Thinned cartilage

Osteoarthritis

Effects of ageing on the respiratory and cardiovascular systems

Respiratory system

Organ	Causes	Result
Lungs	Accumulate damage due to air pollution, smoking, respiratory infections	Lung becomes less elastic and therefore less effective
Ribs	Arthritic changes	Ribcage does not move as freely

Gas exchange is still adequate unless an infection occurs. Respiratory diseases are more common in the elderly:

- **emphysema**
- **bronchitis**
- **pneumonia** – often a complication of influenza; vaccination for 'flu is offered annually to patients aged 65 and over; pneumococcal vaccine is also offered.

Cardiovascular system

Cardiovascular diseases are major cause of death in the UK in men over 45 years and women over 65 years of age.

Organ	What happens?	Causing?	Result
Arteries	Artery walls become less elastic. May show signs of calcification	**Arteriosclerosis** – hardening of the arteries	Loss of elasticity and lessened blood flow Increase in blood pressure
	Fatty deposits, e.g. cholesterol, in artery wall	**Atherosclerosis**	Normal blood flow obstructed; shortage of oxygen-carrying blood at cells. Could result in a stroke
Heart	Exercise – cardiac output reduces	Cannot pump blood as efficiently	Circulation is slowed
	Response of heart is slower and less forcible	Exertion/sudden movement – decrease in cardiac output	Dizziness and loss of balance
	Coronary arteries narrow	Blood supply to cardiac muscle is restricted	**Angina, myocardial infarction** (MI)
	Hypertension – left ventricle works harder	Ventricle may enlarge and become weaker	
	Heart valves become thickened and more rigid	Murmurs	
	Number of pacemaker cells decrease	**Sinoatrial node** beats more slowly	Heart less able to alter its rate

Social consequences of an ageing population

By 2025, it is estimated that those eligible to claim a state pension will greatly outnumber children. This will affect:

- quality and size of workforce
- economy's earning power and productivity
- government tax revenue and expenditure.

We will not be able to rely on the next generation to pay the pensions of the present generation.

There could be an increase in number of very old, frail and dependent people.

QUICK CHECK QUESTIONS

1   State four effects of ageing on our skeletal system.
2   Distinguish between osteoarthritis and osteoporosis.
3   Why do people appear to shrink as they age?
4   What can be done to help in the prevention of osteoporosis?

5   State two effects of ageing on our respiratory system.
6   Distinguish between arteriosclerosis and atherosclerosis.
7   What causes angina?

End-of-unit questions

1 Nail patella syndrome is a rare genetic disease, which causes deformity or absence of some or all of the nails and the absence of the patella (kneecap).
The diagram below shows a pedigree from an affected family. The ABO blood groups of the members of the family are also shown.

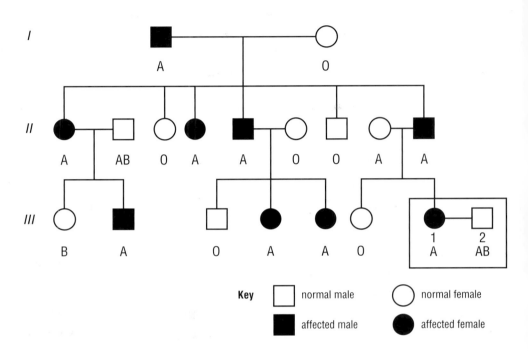

Key   □ normal male   ○ normal female
       ■ affected male   ● affected female

(a) (i)   Using the symbols **N/n** for the nail patella locus and **I^A/I^B/I^O** for the ABO blood group locus, state the genotype of the parents in *generation I*.   (4)

   (ii) State *and* explain the type of inheritance shown by nail patella syndrome.   (3)

(b) The individuals in *generation III*, shown in the box on the diagram, are hoping to have children. They decide to consult a genetic counsellor before they do so. Individual **1** has nail patella syndrome and this condition has occurred several times in her family.
Describe the information that the genetic counsellor could give individuals **1** and **2** about the inheritance of this condition *and* the advice that they might be given before starting a family.
You will gain marks if you use information given in the diagram.   (8)

2 Turner syndrome may be diagnosed while the fetus is still in the uterus, by obtaining cells from the amniotic fluid.
(a) Outline how these cells are treated to diagnose Turner syndrome.   (3)
(b) Why are chromosomal mutations easier to diagnose than gene mutations?   (2)
The improved diagnosis of genetic disease involves some important ethical issues.
(c) Describe *two* **ethical issues** that may be involved.   (4)

3 Many genetic diseases are caused by mutations.
  (a) Define the term 'mutation'. (2)
  (b) A mutation in a somatic cell may not be as serious as a mutation in a
    germ cell. Suggest why this may be the case. (3)
  (c) Many techniques now exist to add or modify DNA sequences. Explain why
    the addition to a cell of a normal gene, leaving the defective gene in position,
    can only work when the disorder results from a recessive allele of a gene. (2)

4 Heart transplants may be recommended for severe coronary heart disease,
  heart failure or for infants born with severe heart defects.
  (a) Outline the ethical issues involved in transplant surgery. (4)
  In addition to the blood groups, it is important that the tissue type should match
  as closely as possible.
  (b) Explain what is meant by the term **tissue type** in this context. (4)

5 A number of drugs and other chemicals affect the action of the nervous system by
  modifying the transmission of nerve impulses at synapses.
  (a) State two functions of synapses. (2)
  (b) When action potentials arrive at the presynaptic membrane, calcium ions
    enter the neurones.
    How do the calcium ions enter the neurones? (2)
  (c) It is known that diamorphine activates inhibitory receptors on the presynaptic
    membrane of sensory neurones in the brain.
    State the *therapeutic* use of diamorphine. (1)
  (d) Contamination from organophosphates in sheep dips, used by farmers to
    kill ticks, may cause illness in the farmers and their families. It is known that
    organophosphates inhibit acetylcholinesterase.
    How do you think organophosphates may inhibit acetylcholinesterase? (3)

6 The diagram below is a longitudinal section through the human eye.
  (a) Label parts A–D. (4)

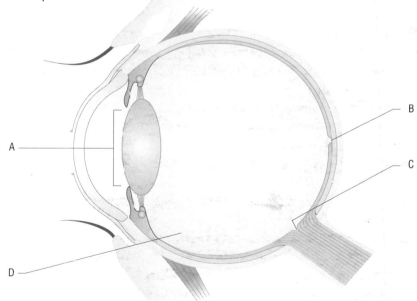

  (b) What are the functions of:
  - optic nerve
  - retina
  - suspensory ligaments? (3)

(c) A schoolboy falls off a rope in the gym. He is taken to hospital and upon arrival he is tested for his level of consciousness by carrying out two eye tests:
  • a pupil response test
  • a blink reflex test.

Explain how these two tests would be carried out in order to find out the boy's level of consciousness. (6)

(d) Optometrists carry out visual acuity tests as part of a routine eye examination. Outline how visual acuity is assessed. (4)

(e) Describe how an action potential is generated when light stimulates a rod cell in the retina of the eye. (6)

7 The diagram below is a longitudinal section through a human brain.

(a) Label parts A–C. (3)

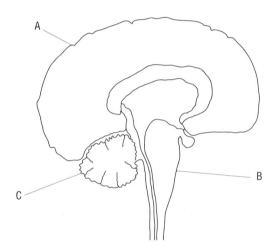

(b) State what is meant by traumatic brain injury. (1)
(c) Describe how an MRI scan can be used to assess brain damage. (4)
(d) An MRI scan could show an area of dead tissue which has been caused by a stroke. State *three* factors that increase the risk of strokes. (3)
(e) Sometimes the body responds to a painful stimulus before the pain is felt; for example, the rapid withdrawal of a hand after touching a hot oven shelf. Explain how this is possible. (4)
(f) The myelin sheath speeds up the transmission of an action potential along the axon.
Explain how the myelin sheath speeds up the transmission of an action potential. (5)
(g) Suggest the likely effects of diseases of the nervous system in which the myelin sheath breaks down. (2)

8 As the prevalence of diabetes mellitus increases patients are encouraged to manage their condition themselves.
(a) State *two* features of the islets of Langerhans which are characteristic of an endocrine gland. (2)
(b) Explain how the islets of Langerhans prevent the blood glucose concentration from falling too low. (3)
(c) If an increase in blood glucose concentration is not corrected for some time, changes may occur in the body that are characteristic of diabetes mellitus. Explain how an excess of blood glucose may cause the signs and symptoms of diabetes mellitus. (6)

9 Maintaining the flow of blood around the body is essential if the organs of the body are to survive and function normally. Some body organs, such as the kidneys, require a particularly high flow of blood.

(a) Explain why the blood flow to the kidneys needs to be high. (2)

(b) Describe the process of **selective reabsorption** in the kidney.
Do *not* include the function of the loop of Henle in your answer. (8)

The kidney produces erythropoietin (EPO) if:
- the oxygen concentration of the blood falls, or
- the volume of the blood plasma increases.

EPO is a hormone that causes the bone marrow to increase its production of red blood cells. Once the oxygen concentration of the blood returns to normal, the production of EPO drops.

(c) Suggest why an increase in blood volume has the same effect on EPO production as a decrease in oxygen concentration. (1)

Although the procedure is banned, some athletes may inject recombinant human EPO (RhEPO) in order to increase their performance.

(d) Explain why injecting RhEPO would increase the performance of an athlete. (1)

(e) Describe what is meant by the term **recombinant erythropoietin**. (2)

10 (a) How does a low body temperature causes the symptoms of hypothermia? (3)

(b) Explain how hypothermia should be treated. (3)

(c) Why is core temperature the most accurate measure of body temperature? (1)

(d) Outline how body temperature is kept at its set point. (6)

11 (a) As a woman ages, she is more likely to have twins than a younger woman. Suggest a reason for this. (1)

(b) Identical twins are produced when one zygote splits into two; non-identical twins are produced from two separate zygotes. Many studies have been conducted comparing non-identical twins with identical twins. Explain why these studies are useful. (3)

(c) Describe *three* signs or symptoms of the menopause. (3)

(d) How does HRT help to relieve the symptoms of the menopause? (2)

(e) Explain how the hormones used in HRT cross cell membranes. (2)

(f) Describe the cyclic and continuous methods of taking HRT. (4)

12 (a) Ageing is one factor which increases the risk of developing Alzheimer's disease. Name *two* other risk factors. (2)

In the UK, life expectancy has shown a steady increase. As the body ages there is a reduction in the efficiency of some organs and tissues.

(b) Outline the effect of ageing on the sense organs *and* peripheral nervous system. (4)

(c) Discuss the **social** consequences of an ageing population to society. (6)

(d) Elderly people may be more at risk from infection. Explain why this is so. (2)

(e) State *one* precaution that should be taken by elderly people to reduce the risk of infection. (1)

13 Alzheimer's disease affects more than 500 000 people in the UK. Despite extensive research into the causes of this disease, there is still no cure.

(a) Describe the changes in the brain associated with Alzheimer's disease. (7)

(b) Why it is difficult to confirm the diagnosis of Alzheimer's disease? (2)

Exam skills: interpreting data

Examiner tip

There are some simple rules for tables that you must follow, when drawing your own table and recording data:

- Tables should be ruled up and have column headings, with units.
- Units must be only in the column headings, not in the body of the table itself.
- The **independent variable** should be in the first column (this is the variable that you change).
- The columns to the right should contain the **dependent variables** (the readings you are taking).

Tables

It is important to have the confidence to describe and interpret data in exam questions. The data may be presented as tables or as graphs. This spread deals with tables.

How to deal with tables in exam questions

Exam questions frequently contain data, which you may not have seen before. You need to engage with this data in order to answer the question well.

Data-handling questions will never be the same, so you cannot learn the mark scheme and hope this will allow you to answer the question. However, there are some steps you should always carry out to help you deal with these types of questions more easily.

Tackling the problem

- A table should have a heading – this tells you what the data are about. So begin by reading this.
- Read the stem of the question. This is the opening paragraph intended to inform and explain some aspects of the question.
- You should always read this opening paragraph to gain good marks. If you do not, you will not have all the information you need. Highlighting key words can help.
- Understand the data by looking at the column and row headings and their units. This gives you some more information.
- Take some time to do this. It is not time wasted. You could start by looking at one of the rows of data and interpreting the figures.
- Perform some simple calculations on the data and manipulate the figures.
- **Describe** the trends and patterns the data show. A quick sketch graph in rough can often help you do this. Here is where you can use data quotes.
- **Explain** the data. A different skill. What do the results mean and why has it happened?
- Quote data figures when answering the question. Once you have understood the data it should be easy to decide what data to use.
- As a general rule, quote pairs of data or comparative points. For example, when quoting the deaths from smoking over a time period you could quote the number of deaths at time X and compare with the number of deaths at time Y.
- Make sure you always include the correct units for the data you quote.
- Decimal places – be guided by the data you have been given as to how many decimal places to quote or to calculate to. Check and see if you have been asked for an answer to the nearest whole number, and numbers of people should always be given in whole numbers.

Worked example

Step 1: Remind yourself that tuberculosis (TB) is an infectious disease that can damage the respiratory system. Name the *type* of organism which causes TB.

This is testing your knowledge of TB. You are not required to know the precise name just the type of organism. Simple questions like this are often the first to be asked.

Step 2: *Read the table title.* Table 1.1 shows the number of *new* cases of TB *notified* to the government each year by two different regions in England. This should help you understand what the data are about.

Step 3: Read the headings in the table. Note these data are per 100 000 in the population, so the data are standardised.

Table 1.1

	Notifications of TB cases/100 000 of population							
	1995	1996	1997	1998	1999	2000	2001	2002
London	30.0	31.8	34.8	35.0	35.4	40.0	40.2	38.6
South West	4.2	4.2	4.4	4.4	4.3	4.6	4.0	4.8

Source: Health Protection Agency Communicable Disease Surveillance Centre, 2002.

Step 4: Describe the general trend for notifications of TB in London between 1995 and 2002.

This is testing your understanding of the data in the table. You may find a rough sketch graph in the space at the bottom of the page helps you see the trends better.

Begin by describing the overall pattern.

Then describe the trend referring to the headings in the table.

Finally refer to any changes in the data. For example in London there is a marked increase in 2000 followed by a decrease in 2002. Support your answer with data quotes.

Step 5: Suggest and explain why the notification rate of TB cases is much higher in London than in the South West of England.

There are two parts to this question. *Suggest* means just that, you are being asked for a suggestion and so there are many possibilities. It shows the examiner that you are thinking about the data. For example, does notification mean the same as disease incidence or could there be other factors that give a higher rate in London? Quote the data here.

Explain requires you to give a *reason* for the notification in London being higher. For this you need to think about any differences between London and the South West.

Step 6: What percentage proportion is there between notifications in London and the South West in 2000? Show your calculations.

This is a simple calculation between the figures notified in the year 2000. Make sure you can do this type of simple maths.

Exam skills: drawing and using graphs

- line graph
- histogram
- bar chart
- line of best fit
- straight line connecting plot to plot

Drawing graphs

A graph allows you to see the trends and patterns more easily than in a table.

Although each graph you are given will not be the same, the skills needed to answer the questions will be similar, so practicing questions with graphs will:
- help in understanding the data
- give help in describing patterns and trends
- help explain the data.

In addition:
- a graph should have an informative title
- the independent variable should be on the y-axis and the dependent variable on the y
- the axes must be correctly scaled and should be labelled with correct units (as a general rule the graph should take up more than half of the available space)
- all the points should be correctly plotted using a cross or a dot with a circle around it
- join the plots with a **straight ruler line** or a *correct* **line of best fit**.

Examiner tip

Draw your graph correctly in the assessed tasks and the extended investigation to avoid losing marks.

Types of graphs

There are three types of graphs useful to you at A level.

Firstly, look at the data and decide which is the independent variable and which the dependent variable.
- A **line graph** is used when the independent and dependent variables are continuous.
- A **bar chart** is used when the independent variable is not numerical, e.g. type of sugar.
- A **histogram** is used when the independent variable falls into numerical categories, e.g. age ranges.

Using graphs

Use these points when describing the graph.
- The title of the graph will tell you about the data, so read the title first.
- It is important to read both the axes and the units to help understand more about the data given.
- Look at the line.
- Use the ruler technique: place a ruler along the y-axis and slowly move it across the graph. Read the data on the line as you move the ruler. Make a note of the general trend line. Note any changes in the trend and where these occur.
- Quote data. A general rule is to use pairs of data where possible and *always* include the units. Two quotes from the x-axis and two from the y-axis are generally needed to describe a change or trend.
- Use a ruler and pencil to draw a line on the graph. This can be from the x-axis to the graph line *or* from the y-axis to the graph line. With this, it is possible to draw a second line from the point indicated on the graph line to the other axis, allowing you to read off the data accurately.

Worked example

If you follow the simple steps and rules given below you will find answering questions with graphs much more straightforward.

Step 1: Read the background information:

The cinnamon spice comes from the bark of the genus *Cinnamomum*. Recently, research has linked the intake of cinnamon to an improvement in type 2 diabetes.

An experiment monitored the effects of cinnamon on glucose concentrations in the blood of 60 people with type 2 diabetes. They were divided into 6 groups of 10 and each group given a different treatment regime over 40 days.

The results are shown in the graph below.

Step 1: Look at the graph lines drawn and *describe* the overall trend. What further processing, if any, is possible with these data? What other information may you need to carry this out?

Step 2: Use the data in the graph to suggest the best treatment to reduce the glucose concentration in the blood.

This requires you to look at the graph and read off and quote the data. Make sure you give units as well.

Step 3: Individual results for mean percentage change in blood glucose concentrations varied about the mean by + or − 2. Explain why the researchers concluded that 'no significant changes in blood glucose concentration were seen in the placebo groups'.

> **Hint**
>
> Use the ruler technique described in the section 'Using graphs' above, to give you the overall trend and any changes.

> **Hint**
>
> A pencil line drawn with a ruler can help you read off the data accurately.

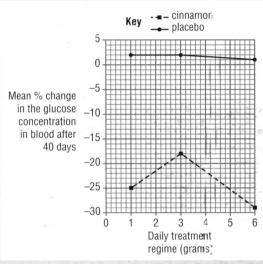

You need to understand statistical conclusions using chance and significance of data. Review your understanding or the conclusions of statistical tests. You may carry out a full statistical test in your investigation or you may be asked to comment on one in a written paper. You must understand what the conclusion means and how to explain it correctly. Remember that a 95% confidence limit is the minimum required. In other words, the probability of seeing this difference by chance is less then 0.05. If the calculation gives a value greater than this 'critical value', then the difference you are seeing in your data is significant at the 95% confidence level.

If vigorous exercise continues for more than a few seconds, anaerobic respiration becomes the dominant form of respiration in the skeletal muscles.

The figure shows the average oxygen consumption by an athlete during and after a race.

Using the information in this figure, explain why there is an oxygen deficit *and* explain the reason for the excess post-exercise oxygen consumption (EPOC). (5)

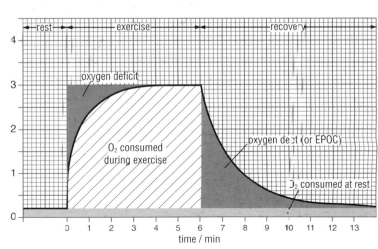

Answers to exam skills questions

Data: worked example

Bacterium/*Mycobacterium tuberculosis*/bacillus;

(Gradually) increases; decrease, after 2001/in 2002; figures to support;

Suggest:
Overcrowding/dense population; poor quality housing/living conditions; more homelessness; greater proportion of drug users; greater proportion of HIV positive; immigration/influx, of people from area with greater TB incidence; arginine vasopressin (AVP); e.g. better diagnosis, qualified reference to air pollution;

Explain:
airborne/droplet infection;

$40.0 - 4.6 = 35.4; \quad \dfrac{35.4 \times 100}{40} = 8.85\%$

(9)

Graphs: worked example

1   Trend shows a small increase in % reduction from 25% at 1 g to 29% at 6 g; however, there is a drop to 18% at 3 g of cinnamon; marked difference compared to placebo; allow data quoted only if % reduction *and* g of cinnamon are quoted with units;

2   Daily cinnamon treatment reduces blood glucose concentration by up to 29%: 6 g of cinnamon gives the largest reduction (29%); 1 g giving 25% reduction; difference is not large between two treatments; 3 g gives the lowest (18%);

3   By chance; statistical tests showed change is not significant/95% likely to be by chance; change in the placebo group *within* normal range of +/*n*2; not statistically significant; fluctuation/differences expected due to differences in individuals;

(9)

Graphs: question

Oxygen deficit:

Difference between oxygen demand and supply; due to an inadequate blood supply to muscles;

EPOC:

Total oxygen consumed after exercise in excess of pre-exercise level; lactate/lactic acid, produced by anaerobic respiration; transported to liver; converted by pyruvate; oxygen needed after exercise to oxidize pyruvate; refer to link reaction; refer to Krebs cycle; use of figs; AVP; e.g. reoxygenating myoglobin/haemoglobin, regenerate, creatine phosphate (CP).

(5)

Answers to quick check questions

Unit 1 – Molecules, blood and gas exchange

Module 1 – Molecules and blood

Blood (page 2)
1 Large numbers of white blood cells suggest infection. Low numbers suggest bone marrow problem/immune problems.
2 B lymphocytes produce antibodies. T lymphocytes destroy diseased and damaged cells.
3 Red blood cells are too large to move through the capillary walls. White cells can change shape to squeeze through. Some white cells produce a chemical that makes the capillary walls more porous.
4 To prevent transmission of infection and disease.
5 The number of blood cells is very high – millions per mm^3 of blood. In order to observe them the cells must be spread very thinly. This also allows better uptake of the stain.

Eukaryotic cells (page 4)
1 Eukaryotic cells contain a nucleus, surrounded by a nuclear envelope, and membrane-bound organelles such as mitochondria.
2 Plant cells have a cellulose cell wall, chloroplasts containing chlorophyll, a permanent large cell vacuole surrounded by a tonoplast.
3 DNA code in the nucleus provides the information. Ribosomes and ER allow the polypeptide to be assembled from the DNA information by joining amino acids. Vesicles transport the polypeptide to the Golgi. Golgi modifies the polypeptide to form the protein. Vesicle transport the modified protein to plasma membrane for secretion or vesicle is held in the cytoplasm for storage.
4 ATP is used to provide the energy needed to drive cell processes.
5 A micrometre is a unit of length. It is equal to 1/1000 of a millimetre.

Membranes (page 6)
1 Phospholipids move easily like a liquid or fluid, and the proteins form the mosaic pattern.
2 Proteins form pores or channels or act as carrier molecules in the membrane. Some proteins in membranes are enzymes catalysing reactions at the cell surface or within organelles. Others act as receptors for chemicals like hormones or as recognition sites on the cell surface membrane.
3 Cholesterol helps stabilise the phospholipid molecules, keeps them fluid and reduces permeability.
4 Microvilli are folds of the cell surface membrane that increase the surface area. Cilia are processes made of protein microfibres that originate from below the cell surface membrane.
5 Membranes within the cell subdivide the cytoplasm and surround organelles. This increases the efficiency of cell functions. They also increase the surface area for metabolic processes. Some membranes isolate harmful substances within the cytoplasm, e.g. lysosomes isolate digestive enzymes. Vesicles are surrounded by membranes and are used for transport within the cytoplasm.

Proteins and haemoglobin (page 8)
1 Diagram to include the basic form of two amino acids, showing an amine group at one end, a carboxyl group at the other end,

and a central region of C, H and R group forming a typical bow-tie shape.
2 Primary structure is the order and arrangement of amino acids joined by peptide bonds to form a chain. Secondary structure is the manner of twisting or folding of the primary structure into an α helix or a β pleated sheet. Tertiary structure is the 3D shape of the globular protein held in place by hydrogen bonds, disulfide bonds, ionic bonds and hydrophobic interactions.
   The quaternary structure of a globular protein is where more than one polypeptide is combined to make one protein, e.g. haemoglobin has four chains combined together. The quaternary shape of the fibrous protein has more than one secondary shape twisted together into a rope shape, e.g. collagen fibres have three polypeptide chains twisted together.
3 The bonds holding the tertiary structure are disrupted. The result is the 3D shape becomes distorted and the protein can no longer carry out its function.
4 See 2 above.
5 Haemoglobin combines readily with oxygen at high ppO$_2$ (high oxygen concentrations) but releases it readily at low ppO$_2$. It is an important protein present in erythrocytes and used in O$_2$ transport in the blood.

Water (page 10)
1 It has areas of δ negative and of δ positive charge. It is attracted to other water molecules and to charged molecules.
2 Water is important in living organisms since most biological reactions occur in water. It is a good solvent since it hydrates many ions and charged solute molecules making it a good transport medium. It is thermostable. Its cohesive properties make it ideal for support and for lubrication. It takes a lot of heat energy to vaporise water so evaporation of water cools.
3 Hepatic vein leaving the liver and the renal artery entering the kidney (relative to the renal vein).
4 Fibrinogen and albumin are large protein molecules in the plasma. They are too large to move through the capillary wall into the tissue fluid.
5 Water potential balance in the blood is maintained by the kidneys' selectively excreting water and salts (remember salts dissociate into ions in solution).

Water potential and diffusion (page 12)
1 Diffusion is the net movement of molecules from an area of high concentration to an area of low concentration (down a concentration gradient).
   Facilitated diffusion is the movement of molecules down a concentration gradient with the aid of a channel or carrier protein.
   Osmosis is a special kind of diffusion where there is a net movement of water molecules down a water potential gradient across a selectively permeable membrane.
   Active transport is where molecules move against their concentration gradient using a carrier protein and ATP.
2 Similarity: both use a carrier protein. Differences: in active transport, ATP is required and the molecules move against their concentration gradient; in facilitated diffusion, molecules move down their concentration gradient and no ATP is needed.
3 Water potential of red blood cell cytoplasm is much lower than pure water and so water moves into the cells down the water potential gradient by osmosis, increasing the volume and rupturing the plasma membrane.

Answers to quick check questions

4   Bulk transport moves large molecules into and out of the cells. The molecules are too large to diffuse through the membranes or to pass through carrier molecules or pores.

Carbohydrates (page 14)

1   Monosaccharides, e.g. glucose, are small molecules with a carbon ring structure. They have side arms of H and OH (hydroxyl) groups. The ratio of hydrogen to oxygen is the same as that in water (H_2O). They are soluble and easily transported within the plasma.
2   Glycosidic bonds are formed by a condensation reaction with the removal of 2H and 1O. These bonds are broken by hydrolysis where 2H and 1O are added. Enzymes catalyse both condensation and hydrolysis reactions.
3   Glycogen is a storage molecule, stored in liver and in muscle cells. It is easily broken down to glucose when needed or synthesised when glucose is in excess.
4   A triglyceride is a large molecule made of a glycerol and three large fatty acid chains. The fatty acids are joined to the glycerol by three ester bonds.
5   A saturated fatty acid has all single carbon–carbon bonds in the hydrocarbon chain. An unsaturated fatty acid chain has fewer H atoms and at least one double carbon–carbon bond, resulting in a bent chain, i.e. a kinky tail.

Enzymes and blood loss (page 16)

1   Fibrinogen, thromboplastin, prothrombin (albumin).
2   Exposed collagen → blood platelets stick and attract other platelets (in presence of Ca^{2+}) → release thromboplastin → acts on prothrombin (and Ca^{2+}) to form thrombin. Thrombin acts on fibrinogen (and Ca^{2+}) → forms fibrin fibres → fibrin forms a mesh which traps blood cells and a clot forms.
3   The active site of the enzyme thromboplastin and the substrate prothrombin have complementary shapes which fit together to form an enzyme–substrate complex. Other proteins do not have the complementary shape and will not fit.
4   High temperatures and pH changes will cause the enzymes and proteins in the blood plasma to denature and the blood cannot be used.
5   The clotting mechanism will not stem severe blood loss, since the fibrin fibres will not hold in place over a large wound, and so no blood cells become trapped and a plug or clot will not form.

Blood transfusions (page 18)

1   Osmotic changes will result in movement of water either into the blood cells causing them to rupture or out of the cells causing them to crenulate (shrivel). In either case the blood spoils. Mannitol and saline (sodium chloride) balance the water potential.
2   Calcium ions act as enzyme cofactor allowing the blood clotting enzymes such as thrombin to function.
3   Correct temperature of 4 °C, correct buffered pH, balanced water potential, removal of Ca^{2+}, and storage of single units in sealed and sterile plastic bags.
4   Health question check; testing for antibodies and viral nucleic acids; blood group testing.
5   Whole blood, leuco-depleted, packed red cells, clotting factors, platelets, plasma.
6   Other blood products are used for specific problems and their treatment. Whole blood is more difficult to store and must be blood group compatible, and so is only used for severe blood loss.

Module 2 – Circulatory and gas exchange systems

The heart (page 20)

1   The heart has two pumps side by side. The left ventricle pumps to the whole body and the right ventricle pumps to the lungs.
2   A single blood cell would enter the heart from the body at the right atrium. From here it moves to the right ventricle and then to the lung via the pulmonary artery. From the lungs it returns to the heart via the pulmonary vein to the left atrium. From the left atrium it moves to the left ventricle and then out of the heart via the aorta to the body.
3   One heartbeat is 0.8 seconds. The rate will be 75 beats per minute. (60/0.8 =75).
4   An ECG is the recording of the electrical impulse as it is transmitted through the heart muscle. An abnormal ECG trace informs of incorrect heart pumping or a malfunction in the contraction of the heart muscle which may occur after a heart attack.
5   Heart rate is the number of heartbeats per minute. A rise in rate will increase the volume of blood pumped out per minute (cardiac output). A decrease will reduce the volume of blood pumped out per minute and so decrease cardiac output.

Circulatory system (page 22)

1   A double circulation increases the efficiency of delivering oxygenated blood to the tissues and allows the systemic blood pressure to be higher than the pulmonary blood pressure. A closed circuit increases efficiency of blood flow since blood is always contained in vessels and will maintain some pressure to continue its circuit.
2   Smooth muscle in the walls of the arterioles contracts to narrow the lumen to reduce blood flow and relaxes to dilate (widen) the lumen to increase blood flow to a tissue or organ. In the skin, arterioles dilate to increase flow to the skin surface and allow loss of excess heat.
3   Capillaries provide a large surface area because a network has large numbers of very small vessels. Walls are one cell thick and the cells making up the walls are squamous (flattened) cells giving a short diffusion distance. Capillary walls are also porous with gaps between cells, and a basement membrane which allows small molecules to cross. Concentration gradient is maintained by the blood flowing onwards.
4   Blood enters the arteries at high pressure causing walls to stretch. Elastic fibres recoil and help maintain the pressure and onward flow. This stretch and recoil gradually evens out the pulsing.
5   Hypertension is raised blood pressure. Values of more than 140/90 mmHg are hypertensive.

The lungs (page 24)

1   A tissue is a group of specialised cells carrying out the same specific function, e.g. ciliated epithelium. An organ is a group of different tissues forming a distinct unit carrying out a specific function, e.g. the stomach.
2   Larger organisms have a small surface area to volume ratio. Tissues lie deep within the body creating a long diffusion distance.
3   Alveoli have a large surface area: a single layer of squamous cells forming a thin layer with a short diffusion distance. They are surrounded by capillaries.
4   Lung has many bronchiole branches. Eachs end in air sacs, which have alveoli to increase the surface area further. Alveoli provide a short diffusion distance between the air and the blood in the capillaries. They maintain a steep concentration gradient with tidal air flow within the lung and blood flow in the capillaries.
5   A peak flow meter measures the rate of air breathed out of the lung. During an asthmatic attack the flow of air out is reduced since the airway is restricted and so the rate will be reduced.

Unit 2 – Growth, development and disease

Module 1 – The developing cell

Mitosis as part of the cell cycle (page 30)

1   They do not undergo mitosis. They divide by binary fission – they do *not* have linear chromosomes, a nucleus, spindle fibres or centrioles.

2   To provide the cellular energy for microtubule formation (protein synthesis), movement of centrioles, movement of chromosomes, formation of new nuclear envelopes, synthesis of new organelles in the next G_1 phase.
NB: chromosomes and centrioles do not move on their own – they are moved by microtubules of the cytoskeleton. These have proteins that act as microtubule motors. Chromatin is condensed into chromosomes for cell division as they are easier to move in that state.

3   Purine molecules consist of two rings/are larger and pyrimidines are smaller/consist of one ring. By a purine always joining with a pyrimidine the rungs of the DNA ladder are the same size.

4   There are also 3000 A and 2000 C. This means there are 10 000 base pairs in this length of DNA. So, there are 10 000 nucleotides and therefore 10 000 deoxyribose groups.

5   (c): the DNA has been replicated throughout S phase so it is doubled by the end of S phase and hence by the beginning of G_2 phase.

Apoptosis (page 32)

1   They are not multicellular. If a cell dies the whole organism has died.

2   If apoptosis is balanced by mitosis then there is no cancer (cell proliferation) and no hypotrophy (degeneration). During growth, mitosis outweighs apoptosis, but this is normal for the growth phase.

3   Analyse the DNA from the cells using gel electrophoresis. If apoptosis has occurred the DNA will have been cut into fragments and a banding pattern will be seen.
NB: You may not learn about electrophoresis until A2, so a question like this may be set in an A2 paper and be a synoptic question. Otherwise you could have a go at answering it and not be expected to give the details of how the DNA is analysed.

4   There are no blebs and vesicles formed that get ingested quickly. Hydrolytic enzymes may be released that damage/digest nearby cells. This would lead to many phagocytes coming to that area.

5   Without ATP, DNA cannot replicate; mitotic spindles can't form so the cell can't divide anymore; nor can it carry out synthesis of proteins/replication of organelles.

Cancer (page 34)

1   They fail to undergo programmed cell death and so continue to divide.

2   The DNA replicates so the cell has to divide to restore the diploid state.

3   They inhibit apoptosis/programmed cell death, and promote uncontrolled mitosis.

4   As only one of the alleles needs to be mutated, it is more likely to be implicated in cancer than other tumour suppressor genes where both alleles need to be mutated.

Stem cells and cell differentiation (page 36)

1   **Pluripotent**: cells capable of differentiating into any one of the 220 or more cell types in the human body/body of organism from which it was obtained.
Multipotent can develop into a limited number of different cell types.

Stem cell: an undifferentiated cell capable of dividing and becoming any type of cell in the organism from which it was obtained.
Differentiation: is the process that stem cells undergo to become specialised or differentiated into a specific cell type. Differentiation involves switching some genes off as the cell matures, whilst ensuring that some are able to be switched on to make the proteins that the differentiated cell needs to perform its specific functions.

2   As drugs are metabolised by the liver, new drugs (medicines) could be tested on such a liver, rather than on animals. It is also possible that such a liver could be used for transplants.

3   Less controversial than using embryos that are destroyed in the process. Some people think it could be the 'slippery slope' to using a cell from an embryo and cloning more embryos. However, such cloning is outlawed in Britain and most countries, and there is no scientific value to doing that as it is the stem cells we need, not embryos. Cloning stem cells is therapeutic cloning.
NB: You would be more likely to be asked a question like this as a synoptic question on an A2 paper.

Cancer – risk factors and methods of detection (page 38)

1   They all involve damage to DNA, which is the genetic material. They involve genes such as oncogenes or changes to tumour suppressor genes. However, these changes are not in the gametes and are not passed on to offspring. There are some cancers that are hereditary, e.g. 5–10% of all breast cancers (the ones that involve the BRAC1 and BRAC2 genes). Some forms of colon cancer are hereditary.

2   Eat at least five portions of fruit and vegetables each day to provide antioxidants and other nutrients that protect against cancer.
Eat oily fish, for vitamin D and fish oils, at least twice a week (NB: fish oils are different from the unsaturated fatty acids in margarine.)
Exercise more; don't smoke; drink only moderate amounts of alcohol.
Use sunscreen (but expose to some sunshine as vitamin D protects against cancer).
Regularly examine own breasts/testes for lumps as early diagnosis always gives better prognosis.
Avoid promiscuity/use condoms to prevent spread of human papilloma virus.
Don't become obese.

3   X-rays are mutagenic and can damage DNA. As DNA damage accumulates, several exposures to low doses may be more harmful than one exposure to a bigger dose. However, very low levels of radiation, such as the background radiation in some parts of the world, seem to have a beneficial effect.

4   They have their own blood supply and blood is 90% water.

Some cancer treatments (page 40)

1   It would be unethical to give cancer patients a placebo as it is not going to improve their condition and cancer treatment has to be given quickly for a good prognosis (outcome).

2   There is social pressure on women to look good, so removal of breast could be traumatic, although she can have reconstruction surgery or use prosthetic breasts. The treatment involves surgery and this costs money. It may be that she would not develop cancer anyway; however, the surgery is cheaper than treating her for breast cancer. If she chooses this treatment and feels more secure that mastectomy will reduce her risk of developing cancer then it would be unethical to withhold it. She may be alleviated of worry; it is not 100% certain to prevent cancer, but reduces the risk greatly.

3   There are many plant species yet to be discovered and some of them may provide useful anti-cancer agents.

183

4 It is the modulation (adjustment) of the immune system for prophylactic or therapeutic treatment of disease.
Therapeutic: BCG vaccine can be used to treat bladder cancer; the antibodies made/T cells stimulated must also work against tumour cells. Using the body's own immune cells, such as antigen-presenting cells primed with tumour cell antigens to stimulate the body's own T killer cells to destroy cancer cells.
Vaccine (prophylactic) against human papilloma virus available to women aged 12–26 years.
Interferon – stimulates T killer cells to kill tumour cells.
Dendritic cells (antigen-presenting cells) taken from patient; exposed to patient's tumour antigens which then, on replacement in body, stimulate T killer cells to destroy (cancer) cells with those antigens.

Module 2 – The developing individual

Meiosis and its significance (page 42)

1 Ovaries and testes.

2 Homologous chromosomes are matching pairs. They are the same size and have the same genes, but not necessarily the same alleles, at the same loci. One (maternal) has come originally from the mother and the other (paternal) has come from the father. Humans have 22 homologous pairs of chromosomes and one non-matching pair – the sex chromosomes (XY). Although the X and Y do not match exactly, there is one part of each that matches with the other, so that they can pair up for meiosis.

3

	Meiosis	Mitosis
Number of divisions	2	1
Products	Four genetically different haploid daughter cells	Two genetically identical daughter cells
Chromosome number	Halved	Maintained
Do bivalents form?	Yes	No
Does crossing over occur?	Yes	No

4 **Prophase 1**: DNA has replicated during interphase and the homologous chromosomes pair up to form bivalents. Each bivalent consist of a maternal and a paternal chromosome. Each chromosome consists of two identical sister chromatids. Crossing over occurs and non-sister chromatids swap alleles. Where they cross over is called a chiasma. The centrioles divide and move to opposite poles and the spindle starts to form as the nuclear envelope begins to break down.
Metaphase 1: Spindle has formed. Homologous pairs of chromosomes, still with their chiasmata, randomly line up on the equator and attach to it by their centromeres.
Anaphase 1: Members of each homologous pair are separated by being pulled to opposite poles as spindle threads shorten.
Telophase 1: Two new nuclear membranes form.
Prophase 2: Spindles replicate and move to opposite poles and new spindles form.
Metaphase 2: Chromosomes attach (randomly assort) to equator of spindle. Each chromosome consists of two chromatids but they are not identical as alleles were swapped in prophase 1.

Anaphase 2: Chromatids separate as they are pulled to opposite poles when spindle threads shorten.
Telophase 2: New nuclear envelopes form around the four haploid nuclei. The cytoplasm cleaves and four haploid cells are formed.

Preconceptual and antenatal care during pregnancy (page 44)

1 May impair healthy development of the sperm or ovum. May damage DNA in sperm or ova.

2 So that they are immune and cannot catch German measles during pregnancy. The virus can cross the placenta and cause birth defects.

3 Her blood sugar levels will often be high and so more sugar will cross the placenta to the fetus.

Diet during pregnancy, and monitoring fetal growth and abnormalities (page 46)

1 A teratogen interferes with gene expression/action. Alcohol is a teratogen and can interfere with normal brain and facial development. This may lead to difficulties with behaviour and attention span, or more severely to fetal alcohol syndrome – impaired growth, facial deformities and severe learning difficulties. Vitamin A is also a teratogen. It interferes with the genes that determine the body plan and so if the early fetus is exposed to too much of it, this can cause birth defects.

2 Carbon monoxide can combine irreversibly with haemoglobin and so reduce the oxygen-carrying capacity of blood. This may happen in maternal and fetal blood so less oxygen is delivered to the respiring tissues of the fetus. The fetus will release less energy from food substrates and this may reduce its rate of growth and lead to low birth weight babies.
Nicotine can constrict blood vessels and reduce the delivery of blood, and hence oxygen, to the fetus. It can also reduce the mother's appetite and lead to malnourishment of the fetus. These reduce the growth and hence the birth weight of the fetus. (A malnourished fetus may develop into an adult who has an increased risk of an early heart attack later in life, and may increase the risk of impairing the health of that person's children.)

3 **(a)** Liver stores vitamin A, so this may expose the fetus to too much of this teratogen.
(b) Soft unpasteurised cheese may contain the bacterium, *Listeria*, which causes mild flu-like symptoms in the mother but can cause miscarriage or stillbirth of the fetus.
(c) Raw eggs may contain *Salmonella* bacteria, which cause food poisoning and can make the mother ill and cause miscarriage, premature labour and stillbirth.

Chromosomal mutations and the use of karyotyping (page 48)

1 If an egg is fertilised by two sperms.

2 It occurred after fertilisation, in one of the cells in an early fetus, during the early mitotic division of the zygote. If the zygote was originally XY but divided unevenly so that one resulting cell had no Y chromosome, then half the cells will be X0 and half XY. This will develop as a Turner's syndrome female. If the zygote was XX but divided so that one cell lost an X chromosome, the descendents of that cell will also be X0.

3 Could cause a healthy fetus to be aborted. Chorionic villus sampling can be carried out earlier in the pregnancy and allow an earlier termination, but carries a greater risk than amniocentesis of causing a miscarriage.

4 Water enters by osmosis, down the water potential gradient, through the partially permeable membrane. There is no cell wall, so the cell swells and eventually bursts as the cell surface membrane ruptures.

Growth patterns during the human life cycle (page 50)

1 Absolute growth rate is the growth (measured by increase in height or mass) divided by the time period. If a child grows 4 cm in 6 months, the absolute growth rate is 0.67 cm month⁻¹. Relative growth rate shows the increase compared to the initial value. If at the beginning of the year a child is 100 cm tall and at the end he is 108 cm tall his relative growth rate is 8/100 = 0.08 or 8%. There are no units as both values in the fraction are in centimetres and so they cancel out.

2 Fetal (from conception to birth); infancy (from birth to about age 12 years); adolescence (age 12–18 years); adulthood.

3 Most girls have their adolescent growth spurt at an earlier age than most boys, as girls generally reach puberty earlier than boys.

4 Base the data on growth in healthy infants who are breast fed. Possibly have different charts for children of different ethnic origins.

Differential growth in humans and maintaining healthy growth (page 52)

1 In order for the head to pass through the birth canal it cannot be any larger, so the brain is underdeveloped. This means babies are helpless and cannot develop the motor skills to walk, talk or fend for themselves for a very long time.

2 Individuals should not be parents until they are emotionally mature as well as physically mature. The first stage of maturity is being able to fully look after oneself and the final stage is being responsible for others. There is no point in the reproductive organs being mature until the person is approaching adulthood and is emotionally able to look after a child. The body's resources need to be invested in developing the nervous and immune systems first, as well as in growth of the whole body to adult size.

3 In a very clean environment the immune system is not exposed to enough pathogens and is not challenged enough. It does not have to respond to worm infestations, and the antibodies normally involved in this response are the ones involved in allergic reactions

4 That we have the genes to make the enzymes needed but that other genes switch them off, except during pregnancy and lactation. Switching them on during pregnancy and lactation ensures that the baby should receive enough of these essential fatty acids.

Module 3 – Infectious disease

Infectious diseases (page 54)

1 (a) A disease that is always present in a particular area.
   (b) A sudden outbreak of a disease that spreads across a country/part of a country/area.
   (c) A sudden outbreak of a disease that spreads across continents/across the whole planet.

2 (a) Both have cytoplasm, ribosomes, DNA, RNA, ATP.
   (b) Prokaryote cell has no membrane-bound organelles, such as mitochondria, endoplasmic reticulum, Golgi apparatus, nucleus. It has smaller ribosomes; it has a cell wall made of murein; many have a capsule; it has a mesosome; some have pili, some have a flagellum/flagella.

3 Lack of protein to provide amino acids to make plasma cell proteins/antibodies so cannot mount an immune response.

4 TB: isolate patients; avoid overcrowding; improve nourishment and living conditions; vaccinate people; educate people to sneeze into a tissue and then burn it; discourage spitting as the bacteria live in dried sputum and then get breathed in; wash hands frequently as viruses in droplets persist on hands and can pass to others.

HIV: make sure all blood for donations is heat treated/screened; issue clean needles for medical use and to drug users; educate people to use condoms; screen pregnant women and discourage breast feeding if necessary.

Antibiotics and superbugs (Page 56)

1 Antibiotics interfere with metabolism (cellular reactions) and viruses do not have any metabolism.

2 A strain/population of a particular bacterium that is resistant to many antibiotics.

3 Isolate all new patients and check to see if they carry MRSA; improve hygiene – hand washing/use of alcohol gels – for all staff, patients and visitors; sterilise any equipment that is used on more than one patient – such as blood pressure monitors; thoroughly clean all wards, equipment, mattresses, beds, curtains, lockers, tables; make staff wear disposable gloves and aprons while examining/treating patients; ban covered lower arms, jewellery, buckles, belts, pens and ties; sterilise patients' skin before an operation – MRSA is easily killed by antiseptics.

Antimicrobials and other medicines from plants (page 58)

1 A substance/chemical that alters metabolism and/or behaviour. Some occur naturally – caffeine in coffee beans and tea leaves, nicotine in tobacco leaves. Some are man-made, e.g paracetamol.

2 Chemicals that kill or inhibit the growth of microorganisms, such as bacteria, fungi and protoctists

3 Many deadly poisons, such as the alkaloids strychnine, brucine and nicotine, come from plants. (However, these may also be used under controlled conditions as medicine, as can morphine.) Many toxins from fungi are very harmful. Disease is a natural phenomena as bacteria are part of the living world (in fact they make up the bulk of the living world but we can't normally see them so we don't realise it). Floods and volcanic eruptions are also natural. All these are harmful. Some man-made drugs are very beneficial, as is surgery, blood transfusion, organ transplantation, vaccination. Genetically engineered (GM) crops provide more food to people in some parts of the world where there is still not enough food.

Vaccination (page 60)

1

	Active immunity	Passive immunity
Antigen presentation	yes	no
Antibodies made by B lymphocytes	yes	no
Memory cells made	yes	no
Time of action	slow	quick
Duration	long/permanent	short/temporary

2 85–95% of the population is immune by vaccination. This means that even if a few cases of an infectious disease appear in the population (perhaps due to new people moving into the area) the chances of an infected person meeting and passing on the disease to a non-immune person are very small. This breaks the chain of infection and stops the spread of the disease/prevents an epidemic.

The immune response (page 62)

1 Infected cells release histamine. This increases blood flow to the infected area (hence it feels hot) and makes capillary walls

more leaky. Tissue fluid seeps out (so the area swells) and macrophages squeeze out into surrounding tissues. Neutrophils and macrophages ingest (phagocytose) and digest the pathogens.

2   T helper lymphocytes produce cytokines that stimulate selected B cells to divide. Cytotoxic T lymphocytes destroy infected cells. B lymphocytes divide by mitosis to become plasma cells or memory cells. Plasma cells produce antibodies. Memory cells remain after the infection giving an immunological memory, making the person immune.

3   Different strains of flu are caused by viruses with slightly different antigens on their surfaces. The memory cells remaining after last year's infection will not recognise the antigens on this year's strain of virus.

Blood groups (page 64)

1   (a) No, as group O blood has anti A antibodies so would agglutinate the donated blood.
    (b) No, as group A blood has anti B antibodies and so would agglutinate the donated blood.
    (c) Yes, as donated blood has no rhesus factor.

2   The fetal blood does not have rhesus factor/D antigen so will not stimulate production of anti D antibodies by the mother.

Tests and vaccines for TB, HIV and cervical cancer (page 66)

1   Because the virus is so unstable and changes its antigens. Memory cells corresponding to one type of antigen will be ineffective against the altered antigens.

2   Some people fear it will encourage early sexual activity/promiscuity. However this is unlikely – promiscuous behaviour and precocious sexual behaviour already exist; fear of HPV has not prevented that. If we have such a vaccine, it is unethical not to use it to prevent cervical cancer, which at best needs drastic treatment and at worst can be fatal. Also, if a girl is not vaccinated she is a source of infection for males who can then infect other females. Prevention is also cheaper than cure/treatment so better for health resources.

3   The Mantoux test. Inoculate TB antigens under the skin, usually on forearm. Examine 48–72 hours later to measure amount of swelling (redness is ignored). If test is positive – person has been exposed to TB and has produced antibodies.

Epidemiology (page 68)

1   Improvement in living conditions, including better nutrition – enabled those infected to make more antibodies/plasma cells; also improved general health; less overcrowding so less likely to spread disease by droplets; better hygiene so less chance of bacteria infecting cuts, food; access to clean water so less water-borne disease such as cholera; sewage disposal and treatment so less contamination of food, reduced risk of faecal–oral route of spreading diseases.

2   There may have been many cases in previous years but during the present year, due to public health measures or people adopting a change of lifestyle, the number of new cases is low.

3   So that any change in pattern of a specific disease can be identified and epidemiologists can try to find out what has caused the change. Is it due to a change in the environment or a change in people's behaviour, or both? Once the causative factors/risk factors have been identified, advice can be given as to how to reduce the spread/incidence of the disease.

The importance and global impact of TB and HIV (page 70)

1   The virus mutates and changes its antigens so there is no vaccine; the disease can be sexually transmitted and it is difficult to get people to change their (sexual) behaviour; there are no obvious symptoms so people often do not know they are infected.

Many countries with a high prevalence of HIV are less economically developed. Their governments have less money to spend on preventive strategies, such as screening blood for transfusions, supplying sterile needles for injections and giving free condoms to people; people are poorly educated so they do not know how to protect themselves or their unborn children from the disease; where there is poverty, more women are forced to work in the sex industry and they can earn more money by having unprotected sex.

Men have to move to cities to find work and when there they may visit prostitutes, become infected and then infect their wives; HIV affects many young adults and this leads to loss of an important section of the workforce.

In areas of civil unrest, rape and deliberate infection with HIV have been used as weapons of war.

Women in LEDCs have low status and men are unwilling to use condoms; people with HIV are often stigmatised and so people may be unwilling to be tested.

2   Some strains of the bacterium are resistant to many antibiotics so the treatment involves taking many antibiotics for a long period. Many people do not adhere to the treatment regime. The vaccine is not very effective, particularly in LEDCs where children are less well nourished; TB spreads by droplets and spreads readily where people live in sub-standard housing; live in overcrowded houses; have a poor diet; drink unpasteurised milk. People may not be educated and therefore do not understand how TB spreads or how to reduce risk of infection; there may be lack of available transport for people to access health-care centres; some countries cannot afford to provide good health care and cannot isolate patients.

Many homeless people suffer from TB and it is harder to access them and give effective medical treatment.

Module 4 – Non-infectious disease

Coronary heart disease (CHD) – 1 (page 72)

1   The heart may be perfectly healthy and functioning normally. The problem is atherosclerosis, i.e. hardened fatty deposits (atheroma) in the arteries. This may acutely (suddenly) reduce or cut off the blood (and oxygen/fatty acids) supply to part of the heart muscle, producing the heart attack.

2   A heart attack is an acute occlusion of a coronary artery so that part of the heart muscle (myocardium) is infarcted – receives no oxygen or nutrients – and cells cannot respire to produce ATP for contraction. This may or may not lead to cardiac arrest, where the heart muscle may fibrillate – beat in an unsynchronised fashion, rather like a fluttering – with no/very reduced cardiac output. Blood cannot circulate to the organs and they fail, causing death

3   (a) A defibrillator sends a large voltage through the heart muscle. This stops it fluttering and when it starts to beat again it will hopefully beat in synchronisation.
    (b) Aspirin is an anti-inflammatory. It inhibits a metabolic pathway that can lead to a blood clot (thrombus) forming. It is also very cheap and has very few (none for most people) side effects.

Coronary heart disease (CHD) – 2 (page 74)

1   Get your blood pressure checked as there are no symptoms with high blood pressure but it is a big risk factor for CHD. Stop smoking as this is also a big risk factor. Drink alcohol in moderation: 2 units per day with 2 alcohol-free days each week is much better than binge-drinking at weekends. Also make sure you are taking exercise and eating a healthy balanced diet with not much saturated fat and not much salt. Replace fatty and processed food with fresh fruit and vegetables and eat less red meat and more oily fish.

2 It is more prevalent in affluent countries but within those countries is more prevalent/has a higher incidence among people less well off.

3 Some people may be heavy due to having a lot of muscle, which is denser than fat. Hence rugby players may appear to be overweight according to their BMI. Also, carrying a lot of fat in the body cavity, around the waist area, greatly increases the risk of heart disease.

Lung disease (page 76)

1 Lung cancer is slow to develop. It takes about 30 years before symptoms appear and before the tumour is large enough to show up on an X-ray or scan. The huge increase in cases of lung cancer in men in the UK in the late 1940s occurred 30 years after smoking became very popular with men in the UK.

2 Elastase digests elastin protein in the alveoli. They can stretch on inhalation but can't recoil so some burst. This reduces the surface area of the lungs for gaseous exchange. Less oxygen combines with haemoglobin in red blood cells. Less oxygen reaches respiring muscle cells so they cannot make much ATP for contraction.

3 (a) When having an attack.
(b) Each morning, or twice a day, to prevent an attack occurring.

Types and distribution of diabetes (page 78)

1 It causes many complications and an increase in the number of people suffering from CHD and strokes, as well as blindness, kidney failure and gangrene. This costs a large amount of money (far more than is spent on cancer treatments), and so will be a huge drain on the health service in terms of both money and personnel. This costs the government a lot of money. It also means loss of economic productivity for the nation and more disability payments within the population.

2 Atherosclerosis develops therefore increased risk of CHD and stroke; blindness; kidney failure; chronic infections; wounds fail to heal and ulcers develop; circulatory problems leading to gangrene and loss of feet or lower legs; inflammation of nerve endings (neuropathy); reduces life expectancy by 10 years; ketoacidosis and coma which could lead to death.

3 Eat a balanced diet; three meals a day and no snacks. Don't eat processed food/junk food as it contains a lot of fat, sugar and salt. Reduce fat intake; reduce intake of sugary foods and drinks. Eat more fresh fruit and vegetables. Take more exercise, don't become overweight.

Diagnosing type 2 diabetes and monitoring blood glucose levels (page 80)

1 In healthy people there may be small amounts of glucose in urine and the test strips do not accurately measure how much glucose is present.

2 Carried out early in the morning, after the patient has fasted (not eaten anything) for at least 8 hours; patient given a drink containing 75 g glucose; blood taken after 2 hours and tested for glucose level. If the level of glucose in the blood is higher than 6 mmol cm^{-3} then there is impaired glucose tolerance. If the level is above 7 mmol dm^{-3} then the patient may have diabetes. Further tests will need to be done to make sure.

3 Place a test strip in the biosensor. Clean the fingertip with alcohol and use a lancet to make a small prick in the fingertip. Squeeze a drop of blood onto the test strip. After about 30 seconds, note the reading. In the biosensor strip is an enzyme that converts glucose to gluconolactone. A small electric current is produced. An electrode in the test strip converts this current to a numerical value. This is displayed digitally on a screen to show the blood glucose concentration.

The biosensor has a memory so results over a period of time can be kept and compared. A person with diabetes should

try to make sure that their blood glucose concentration is 4–7 mmol dm^{-3} before meals, less than 10 mmol dm^{-3} 90 minutes after a meal, about 8 mmol dm^{-3} at bedtime.

Unit 4 – Energy, reproduction and populations

Module 1 – Energy and respiration

Respiration (page 88)

1 ATP is easily made from ADP and phosphate when sufficient energy is available. It is also easily broken down to ADP and phosphate to release energy as it is needed. It is an intermediary molecule between energy-releasing molecules and energy-consuming reactions.

2 Substrate level phosphorylation: sufficient energy released from conversion of one substrate to another to make ATP. Oxidative phosphorylation: ATP formed as electrons pass through a series of carrier molecules in the electron transport chain. O_2 is the final acceptor. Reduced NAD and FAD are the source of the high energy electrons. Photophosphorylation: ATP formed when electrons excited by light energy pass through a series of electron carriers. In ALL three: ATP is made by a condensation reaction between phosphate groups and ADP, producing ATP plus water.

3 Coenzyme A is required to activate acetate and allow it to combine with oxaloacetate and begin the Krebs cycle.

4 One ATP from substrate level phosphorylation, reduced FAD and three reduced NAD, plus two molecules of CO_2.

5 In redox reactions there is alternative reduction as electron is accepted and oxidation as it is passed on. Each of the carrier molecules in the electron transport chain is reduced and then oxidised as the electron passes along the chain of carriers.

Chemiosmosis (page 90)

1 Mitochondria contain enzymes in the stroma allowing Krebs cycle to proceed. The inner membranes (cristae) provide a large surface area and contain carrier molecules and enzymes for the electron transport chain plus ATP synthase (the stalked particles). The outer membrane is impermeable to hydrogen ions so a gradient can build up in the intermembranous space.

2 H atoms from reduced NAD enter the chain. They split into protons and electrons. The electrons continue to pass from carrier molecule to carrier molecule along the chain. The protons are actively pumped out into the intermembrane space creating an electrochemical gradient. The protons flow down this gradient through ATP synthase generating 1 ATP. This occurs three times along the chain so each H can generate 3 ATPs.

3 NAD is an H acceptor. It is vital to remove H as formed to allow glycolysis to continue. In aerobic respiration this occurs since the electron transport chain moves the H from NAD onto the next carrier in the chain. In anaerobic respiration this does not occur, and so the NAD must be recycled to continue accepting H as they are released, so some ATP can be produced from glycolysis.

4 The organic molecule is used in respiration either at the start of glycolysis (glucose) or later in the pathway by conversion to acetyl CoA (lipids). Glucose is the main respiratory substrate.

5 More than one substrate is usually respired. There is usually a mixture leading to intermediate values. Gas leakages in the respirometer may give inaccurate readings. Temperatures may vary leading to expansion of gases giving inaccurate readings.

6 The RQ is determined by measuring the volume of CO_2 released and dividing by the volume of O_2 consumed.

Answers to quick check questions

Exercise (page 92)

1   To prevent straining the heart and putting it under too much pressure; this is the best level to ensure maximum benefit and most exercise will still be aerobic.
2   Nitric oxide secretion causes vasodilation and increases blood flow to muscles and back to the heart. This increases stroke volume at systole.
3   Increases capillaries and muscle fibres, increases the number of mitochondria in the muscle, increases the stores of glycogen and myoglobin, increases heart muscle, reduces resting heart rate and blood pressure, gives a higher VO_2max and vital capacity, and higher maximum breathing rate.
4   Illegal enhancement using steroids increases muscle bulk and protein synthesis putting a strain on the liver, kidney and heart. Blood doping increases the work load of the heart and increases the strain. It also increases the work load on the liver and kidney.
5   Recombinant erythropoietin increases the volume of O_2 carried in the blood as more red blood cells are formed. This increases aerobic fitness.

Making haemoglobin (page 94)

1   DNA code → copied (transcription) → forms mRNA → introns are edited out by enzymes as the ribosomes hold in place → edited mRNA used as template to make polypeptide as tRNA brings correct amino acid into place → ribosome holds the amino acids in place as peptide bonds are formed to create a polypeptide chain (translation).
2   RNA has three forms: mRNA transcribes the DNA code and carries this to the ribosome; tRNA attaches correct amino acid and brings it to the complementary codon on mRNA. rRNA (ribosomal) holds tRNA and amino acid in place while peptide bonds form between adjacent amino acids.
3   Hypoxia is shortage of oxygen delivered to cells. This reduces aerobic respiration so fewer ATP molecules are formed.
4   DNA repair: DNA polymerase checks for DNA mistakes; restriction enzymes cut out the DNA mistakes; ligases repair the damage. Recombinant DNA: restriction enzymes cut out the DNA gene code identified by researchers; the same enzymes cut into the receiving DNA which may be a bacterial plasmid; ligase enzymes splice the DNA gene into position.
5   Telomerase repairs telomeres so cell division can continue. Cancer cells express telomerase so can carry on dividing.
6   Apoptosis is programmed cell death. It occurs when DNA is no longer repairable. Removal of damaged cells or cells with damaged DNA help to keep tissues functioning correctly.

Haemoglobin and myoglobin (page 96)

1   Haemoglobin (Hb) is found in red blood cells, and is a complex 3D protein of four polypeptide chains each with a haem group. It transports oxygen. Myoglobin is a store of oxygen in muscle cells; it is a simple 3D protein with one polypeptide chain and one haem group.
2   Oxyhaemoglobin is Hb combined reversibly with four O_2 molecules.
3   The curve shows how Hb loads with O_2 when pO_2 is high as in the lungs and only releases it when pO_2 is low as in the respiring tissues.
4   CO_2 reacts with water using the enzyme carbonic anhydrase and forms carbonic acid which quickly dissociates into H^+ and hydrogencarbonate ions. The H^+ forms HHb (haemoglobinic acid) as the Hb buffers it and displaces the O_2 from Hb.
5   Myoglobin is the O_2 store in muscle cells where it forms oxymyoglobin. Creatine phosphate is a reserve energy store used to restore depleted ATP reserves.
6   EPOC is the total O_2 consumed after exercise less the resting level of O_2 consumed. Read the volume under the graph line after exercise has stopped and subtract the resting volume of O_2 needed.

Skeletal muscle (page 98)

1   A muscle fibre is a single muscle cell, usually bundled together to form a muscle. Microfibrils are the protein fibres within the muscle sarcoplasm. There are two types in muscle cells – actin and myosin microfibrils.
2   A sarcomere is a unit within the muscle fibre made up of bundles of actin and myosin in a regular pattern.
3   The hydrolysis of ATP is required in muscle contraction to release the myosin head and return it to its original position.
4   Calcium ions are released when a nerve impulse reaches the sarcoplasmic reticulum. Calcium ions bind to the troponin and remove the blocking molecule in the binding site on the actin. This site is now free for the myosin head to attach and begin the rowing action.
5   The power stroke rows the actin in (like a rope being pulled in by a boat man) so the fibres slide past each other. This causes the whole sarcomere to shorten as the actin and myosin slide past each other.
6   When the creatine phosphate is used up, the muscles' energy store has been used. Any oxygen stored as oxymyoglobin is utilised in aerobic respiration, and finally the contractions continue using ATP generated anaerobically. As lactate builds up, even this process is inhibited. Unless sufficient oxygen reaches the cell, the myosin head can stay attached to the actin if ATP levels fall, leading to 'cramp'.

Module 2 – Human reproduction and populations

Human reproductive systems (page 100)

1   The sex organs produce a number of hormones that are released directly into the bloodstream.
    Human reproductive systems
2   In female: produce female gametes; location of fertilisation; location of development of fetus; produce female sex hormones. In male: produce male gametes; produce male sex hormone; introduce the sperm into the female tract for fertilisation.
3   The female tract has a uterus where the fetus develops and is protected until birth. The two fallopian tubes carry the sperm up towards the ovaries to allow fertilisation with the oocyte soon after ovulation.
    The male tract has an extended sperm duct and urethra to allow sperm to be passed into the female's body, without large amounts of waste into the environment.
4   The follicle cells are specialised cells that are part of the ovary. They produce sex hormones and form a fluid-filled cavity as they enlarge. The egg cell is another name for the oocyte and is the female sex cell within the follicle cavity.

Hormones, gametogenesis and fertilisation (page 102)

1   In the fetus, diploid germ cells divide to form oogonia. Meiotic cell division begins and stops at prophase 1 forming primary oocytes. After puberty, one oocyte continues meiosis to metaphase 2, now forming the secondary oocyte. This is released at ovulation.
2   The acrosome releases digestive enzymes on contact with the zona pellucida.
3   Gonadotrophin → stimulates release of FSH from anterior pituitary. FSH → binds to follicle cells in ovary → stimulates follicle to mature → oestrogen is released from mature follicle → rising levels inhibit FSH which stimulates LH surge (after a few days). LH → stimulates ovulation and development of corpus luteum. Oestrogen → causes uterus lining to thicken. Corpus luteum → releases progesterone → rising progesterone levels inhibit FSH, and maintain and vascularise the endometrium causing the production of glycogen-rich mucus.

4   Fertilisation begins when the sperm makes contact with the secondary oocyte. Acrosome releases enzymes which digest a pathway through the zona pellucida. One sperm enters and binds to a receptor on the oocyte membrane to stop other sperm entering. The secondary oocyte completes meiosis and the male nucleus fuses with the remaining oocyte nucleus (one oocyte forms a polar body), and a diploid zygote forms. Fertilisation is vital to restore the chromosome number to diploid and is one source of variation.

Pregnancy and contraception (Page 104)

1   The blastocyst is the hollow ball of dividing cells initially formed from the zygote. The trophoblast forms from the outer layer as the ball sinks into the uterus lining and processes begin to develop around the outside.

2   HCG is produced from the chorion of the developing embryo. It stimulates the corpus luteum and prevents it shrivelling so oestrogen and progesterone production is continued and the uterus lining is maintained (vital for the pregnancy to continue). Pregnancy tests are designed to detect HCG by using antibodies that attach to the HCG molecules. If it is present there must be a pregnancy since it is only produced by the developing embryo.

3   Oxytocin starts the birth process, stimulates uterine contractions to keep birth continuing, and after birth acts as a releaser hormone as the baby suckles to stimulate milk release.

4   Positive feedback mechanisms produce an increased effect as the action is detected. So, as birth contractions occur and are detected this stimulates more oxytocin to be released, and so contractions increase in number and intensity. This is an unusual mechanism as most hormones are controlled by negative feedback.

5   Discussion should include reference to health risks such as increased risk of thrombosis and breast cancer, and increased chance of sexually transmitted diseases. Discuss whether use of the IUD or morning-after pill constitutes an abortion since the fertilised embryo is prevented from implanting. Discussion could also include the need to limit family size for social or economic reasons.

Infertility (page 106)

1   Artificial FSH stimulates follicles to mature if given at the correct time, and artificial LH given a few days later continues the female cycle and triggers ovulation. GnRh may be given to stimulate FSH from the pituitary, or other hormones may stimulate GnRH.

2   IUI is injection of semen near the fallopian tube within the uterus. ICI is injection of sperm near the cervix.

3   IVF is in vitro fertilisation where sperm and oocytes are mixed outside the body. Embryos are then selected and placed into the uterus a few days later. GIFT involves sperm and oocytes passed directly into the fallopian tube. It may involve a donated oocyte. ZIFT is the formed zygote passed into the fallopian tube. ICSI (intracytoplasmic sperm injection) injects the sperm directly into the oocyte to ensure fertilisation.

4   The law which allows male donors to be identified has resulted in fears that donations may reduce, since donors are concerned that they may find themselves required to assist financially or emotionally with the offspring. This is a misconception.

5   Multiple pregnancies have an increased risk of miscarriage, early birth and low birth weight. Also, pre-eclampsia in the mother. 'Vanishing twin' syndrome can lead to additional risks.

6   Monoclonal antibodies can be used to detect HCG in the mother's urine. A small stick with the antibodies attached is dipped into an early morning urine sample. Any HCG binds to the antibodies on the stick. The antibody/HCG complex will now bind to coloured beads on the stick and form a complex of HCG, antibody and bead. The urine seeps up the stick carrying the complex with it. When it reaches fixed antibodies higher up the stick these will bind and give a coloured band if they are present. If they are not present, no coloured band forms in the test window. A positive control gives a blue band to demonstrate that the chemicals are all functioning.

Photosynthesis and respiration (page 108)

1   Respiration releases carbon dioxide as a waste product, and photosynthesis uses carbon dioxide to fix ribulose bisphosphate as part of the build up of carbohydrates and other food molecules. Remember that plants respire AND photosynthesise in the light. In the dark, they just respire.

2   Chloroplasts have an outer membrane and a series of highly folded inner membranes containing the photosystem pigments such as chlorophyll and also electron carrier molecules. The membrane stacks increase the surface area and are positioned to maximise light absorbance. The matrix contains enzymes needed for the light-independent stage such as Rubisco.

3   ATP and reduced NADP.

4   Because ribulose bisphosphate is regenerated each time to begin the cycle again.

5   Rubisco is the enzyme catalysing the reaction which 'fixes' carbon dioxide to ribulose bisphosphate leading to the formation of GP.

6   CO_2 combines with ribulose bisphosphate and forms a 6C molecule. This in turn is the basis of all the macromolecules such as lipids and carbohydrates which are passed up the food chain as organisms are eaten.

Cycles (page 110)

1   Microorganisms act as decomposers. Nitrogen in the air is fixed by Rhizobium. Nitrosomonas oxidises ammonium to nitrites and Nitrobacter oxidises nitrites to nitrates, a form more useful for uptake by plants. These are nitrifying bacteria.

2   Less energy is passed on at each level since some energy is trapped in the structural molecules in tissues at that level, some of which may be inedible. Some energy is lost as heat in respiration, in movement and in metabolic waste.

3   $2500 - (800 + 1580) = 120$. $120/2500 \times 100 = 4.8\%$

4   Some animals are more efficient in converting energy into new material, e.g. rabbits are more efficient as are impala. Less energy is lost in transfer from one level to the next.

5   Respiration, photosynthesis, decomposition and combustion.

Food production and the ecosystem (page 112)

1   Advantages: Extensive farming has low maintenance costs and is less damaging to the environment. Intensive farming has a high yield which is beneficial for the increasing population and the cost of food can be kept low. Disadvantages: Extensive farming has low yield since there is little added fertiliser and it can support few live stock. Intensive farming uses excessive levels of fertiliser and pesticide, harming the environment and increasing levels of environmental damage. There is high energy demand to drive the machinery and biodiversity is reduced.

2   Intensive farming may cause fertilisers and pesticides to be washed into the water table, contaminating drinking water and rivers and streams. Eutrophication is one environmental hazard due to excess fertiliser. Biodiversity is reduced leading to extinction of various species. Loss of biodiversity can in turn damage crop yields as pollinating insect numbers may fall. The environment may become 'impoverished', with fewer species of birds, animals and plants, and hence less attractive. Finally, some plants may disappear which have potential use as sources of medicinal compounds.

3   New dunes form at the sea edge of the dunes; they are barren. Primary colonisers begin the succession with plants such as marram grass which hold the dune in place with their roots.

The dune has now changed providing more nutrients with more species colonising. This makes the original colonisers unsuitable to survive and they die off. The humus content increases, animals begin to colonise as do decomposers. The humus content increases further. Finally, the climax community appears as the best suited to the environment. Grazing can 'deflect' the succession leading to a plagio-climax.

4   Medical advances mean that more infants survive into childhood and adulthood. More old people survive as do the really ill. These points cause the population to rise since there is a lower death rate. Agriculture has established monoculture as being the most efficient, but it decreases the habitats for insects vital for pollination (e.g. bees) and increases the risk of disease spreading. Increased output from farms means more human survivors at the expense of human populations.

5   Carbon footprint may be reduced by avoiding or reducing air travel, reducing car travel, sharing vehicles rather than driving numerous separate trips. Turning off electrical appliances when they are not needed, re-using bags and reducing 'food miles' can also help.

Unit 5 – Genetics, control and ageing

Module 1 – Genetics in the 21st century

Inheritance of human genetic disease (page 120)

1   One of the different forms of a gene which occupy the same locus on homologous chromosomes.
2   Phenotype
3   An alteration of DNA sequence.
4   The secretion of abnormally thick, sticky mucus which obstructs the airways and blocks the secretion of digestive enzymes from the pancreas and the transport of sperm.
5   A mutation results in the triplet CAG being repeated over and over resulting in Huntington proteins with variable numbers of glutamine residues.
6   When both alleles affect the phenotype.
7   $I^A I^B$

Sickle cell anaemia and haemophilia (page 122)

1   $Hb^S Hb^S$
2   Codominant alleles both affect the phenotype.
3   Likely to have immunity to the malaria parasite.
4   Blood smear – look for sickle-shaped erythrocytes.
5   Y
6   A normal man ($X^H Y$) and a carrier woman ($X^H X^h$) have children:
parental genotypes $X^H Y$ $X^H X^h$
gametes X^H Y X^H X^h

	X^H	Y
X^H	$X^H X^H$ normal female	$X^H Y$ normal male
X^h	$X^H X^h$ carrier female	$X^h Y$ haemophiliac male

The use of pedigree diagrams (page 124)

1   To predict the risks of a particular couple having a child who might suffer from a particular genetic disease, and to determine the mode of inheritance (dominant, recessive, etc.).
2   (a) 10 + 1 possible, (b) Alice and Beatrice, (c) sex-linked and recessive.
3   When genes nearby on the same chromosome tend to stay together during the formation of gametes.

4   The frequency of crossovers between particular gene loci.
5   A to C 21; C to B 9
Each 1% crossover = 1 map unit. If A to C has 30% crossover then they are 30 map units apart. C to B has 9% crossover so B is 9 units from C. B to A has 21% crossover so B is 21 units from A. This means that the order must be: A to B (21 units), B to C (9) with 30 units between A and C.
e.g. A 21   B 9   C (30)

Non-disjunction and translocation (page 126)

1   Non-disjunction
2   A piece of chromosome breaks off and is transferred to another chromosome.
3   X
4   Males
5   (1) **Regular trisomy 21** – all the cells have an extra chromosome 21; (2) **mosaic** – only some of the cells have an extra chromosome 21; (3) result of **translocation** when the end of the long arm of chromosome 21 joins another chromosome.
6   Down's syndrome.

Gene technology (page 128)

1   A bacterial enzyme which cuts DNA.
2   Blunt ends do not have single chains of unpaired nucleotides as in sticky ends.
3   The shorter the length of the DNA fragment, the further it travels in a set time.
4   DNA fragments are negatively charged.
5   Denaturation, annealing, synthesis (elongation).

Genetic engineering in microorganisms (page 130)

1   To transport the DNA into the host cell.
2   Because in eukaryotic cells genes need to be 'edited' after transcription and the polypeptide may need modifying after translation in the Golgi apparatus. These processes would not necessarily happen in prokaryotic cells so a functional protein would not be made.
3   In milk or urine.
4   An experimental technique that uses genes to treat or prevent disease.
5   Somatic and germ cells.
6   Will result in permanent changes that are passed down to subsequent generations.

Human Genome Project and genetic counselling (page 132)

1   By electrophoresis.
2   To find genes associated with genetic conditions; genetic tests to enable patients to learn their genetic risks for disease; may start an age of personalised medicine where an individual's genetic code can be used to prevent, diagnose and treat diseases.
3   1 in 4
4   Genetic counsellors work with people concerned about the risk of an inherited disease. They explain the risks of passing on a genetic disorder to the next generation and/or choices available to people.

Transplant surgery and cloning (page 134)

1   Compatibility is the ability to accept transplanted tissue. Ideally, the compatibility should be 100% (6 out of 6) so that the risk of rejection is eliminated.
2   The area of chromosome 6 where the loci for the HLA antigens are situated.
3   Cadavers and living donors.

4   Immunological rejection and the risk of catching a disease communicable from animals to humans; particular concern about retroviruses.

5   By using blastocysts that are left over from *in vitro* fertilisation; by cloning embryos and harvesting stem cells from them and from aborted fetuses.

6   Renewable source of replacement cells and tissues; to test new drugs; to make 'designer babies'.

Module 2 – The nervous system
Visual function (page 136)

1   The central nervous system (CNS) is the brain and spinal cord. The peripheral nervous system consists of the nerves between the CNS and the rest of the body.

2   The iris

3   To focus rays of light onto the retina.

4   Cones

5   Ganglion cells.

The response of sensory receptors to a light stimulus (page 138)

1   The rhodopsin changes shape – 11 *cis*-retinal has a kink at carbon 11; *trans*-retinal has no kink.

2   Depolarised

3   Snellen

4   By looking at a collection of cards made up of different coloured spots which are arranged so that those of a particular colour form images or numbers.

5   It indicates whether there is damage to the optic nerve or the brain. It can also indicate that the person's nervous system is affected by alcohol or other drugs.

6   The rapid closing of the eyelid when something threatening approaches the eye.

The brain and autonomic nervous system (page 140)

1   Helps to protect and cushion the brain.

2   Medulla oblongata

3   Cerebrum (forebrain)

4   Wernicke's area in the temporal lobe of the cerebrum.

5   Control of speech, memory, intelligence, emotion; concerned with vision, hearing and integration of senses.

6   In the medulla oblongata.

7   Peripheral nervous system transmits impulses from CNS to skeletal muscles (voluntary muscles), whereas autonomic nervous system transmits impulses from CNS to involuntary muscles and glands.

Neurones (page 142)

1   Wrapped around the axon or dendron. It acts as an insulator so speeds up the transmission of a nerve impulse.

2   Sensory neurone transmits impulses from a receptor to the CNS whereas a motor neurone transmits impulses from the CNS to an effector.

3   To transmit nerve impulses between other neurones; for example, between a sensory and a motor neurone.

4   Axon carries the impulse away from the cell body of a neurone whereas a dendron transmits the impulse towards the cell body.

5   The response is rapid because the decision-making processes of the brain are not involved and the neurone pathway is short.

6   Receptor → sensory neurone → relay neurone in spinal cord → motor neurone → effector

Transmission of a nerve impulse (page 144)

1   The potential difference or voltage across the neurone cell membrane while the neurone is at rest. The inside is negative compared to the outside (at about −70 millivolts).

2   Depolarisation: when stimulated past threshold the sodium channels open and sodium rushes into the axon, causing a region of positive charge within the axon – this is depolarisation. Hyperpolarisation is when the polarity becomes restored and there is a slight 'overshoot' in the movement of potassium ions (−70 to −90 mV).

3   If the depolarisation of the membrane does not reach the value of the threshold potential (about −50 mV) then no action potential is created.

4   An action potential is transmitted along a neurone due to the movement of sodium ions. When stimulated past threshold sodium channels open and sodium ions rush into the axon, causing a region of positive charge inside. The region of positive charge causes nearby voltage-gated sodium channels to close and potassium channels to open wide so potassium exits the axon. Local circuits are set up between the depolarised region and the region on either side of it. Sodium ions flow sideways inside the axon, away from the positively charged region towards the negatively charged regions on either side. The axon membrane ahead depolarises, generating an action potential. This process continues as a chain reaction along the axon. The influx of sodium depolarises the axon, and the outflow of potassium repolarises the axon.

5   It means that action potentials only move in one direction along the axon.

6   Saltatory conduction occurs when an action potential jumps the large distances from node to node (nodes of Ranvier – in myelinated neurones the voltage-gated ion channels are found only at the nodes of Ranvier; between the nodes, the myelin sheath acts as a good electrical insulator preventing ions moving in or out). Speeds up the transmission of action potentials.

Synapses and brain injury (page 146)

1   A junction between one neurone and another neurone or effector.

2   A chemical that diffuses across the cleft in a synapse to transmit a signal to the postsynaptic neurone.

3   Synapses in which the neurotransmitter is acetylcholine are called cholinergic synapses.

4   To release energy to move the synaptic vesicles to the presynaptic membrane.

5   Action potential arrives at the synaptic bulb; the voltage-gated calcium channels open and calcium ions (Ca^{2+}) diffuse into the bulb; this causes the synaptic vesicles containing ACh to move and fuse with the presynaptic membrane. ACh is released into the synaptic cleft, and diffuses across to bind with receptors on the postsynaptic membrane; sodium ion channels open in the membrane and sodium ions enter the postsynaptic neurone, depolarising the membrane and initiating an action potential; if the threshold is reached, the enzyme acetylcholinesterase hydrolyses the neurotransmitter ACh.

6   Synaptic vesicles are confined to the presynaptic side of the cleft, and protein receptors to the postsynaptic side.

7   An injury caused by the head being hit by something or shaken violently.

Brain and spinal cord damage (page 148)

1   CT scan, MRI scan, nerve conduction velocity test.

2   Because of the formation of scar tissue which produces inhibitory molecules; poor regenerative response.

3   Pluripotent: a stem cell that has the potential to differentiate into almost any cell in the body.

4   A stroke has a detrimental effect on short-term memory (a working memory which stores information about recent happenings), but long-term memory (the store of autobiographical information, language, acquired knowledge) is usually well preserved in stroke victims.

5   Two of the following:
    show the patient photographs and help them identify people they used to know; show photographs in magazines or newspapers and help them identify some of the well-known people featured; play a game with the patient in which they are shown a tray of objects – cover the tray and ask them to name as many of the objects as possible; have a large clear calendar in every room so that they can keep track of the day; leave clear reminder notes around the house so that the patient knows what they should be doing and when.

Drugs (page 150)

1   Any chemical that affects the way your body functions; how you think and feel.
2   Dopamine: does not get into the brain, as cannot pass the blood–brain barrier.
3   Diamorphine
4   They combine with the opioid receptors in the brain and block the transmission of pain signals.
5   Therapeutic – a mild analgesic and sedative which may relieve the symptoms of multiple sclerosis, hypoglycaemia and other disorders; it has been used as a medication for the terminally ill when other treatments have failed to relieve distress. Recreational – most common effects are talkativeness, cheerfulness, relaxation, and greater appreciation of sound and colour; it usually helps to relax the user.
6   Physical dependence – can change the body chemistry so that if someone does not get a repeat dose they suffer physical withdrawal symptoms. Psychological dependence is more common – people use the drug experience as a way of coping with the world.

Module 3 – Homeostasis

The importance of homeostasis (page 152)

1   Homeostasis is the maintenance of a stable internal environment.
2   Because living cells can function only within a narrow range of oxygen and nutrient availability, temperature, pH, and ion concentration. These fluctuate with changes in the environment.
3   A receptor receives information; an effector brings about a response.
4   Negative feedback occurs when the body senses an internal change and then activates mechanisms that reverse the change.
5   Autonomic nervous system is responsible for homeostasis. These controls are done automatically below the conscious level.
6   Core body temperature
7   Vasoconstriction of skin arterioles, contraction of hair erector muscles, shivering, decrease of sweating.
8   In the hypothalamus.

Control of temperature (page 154)

1   More blood flows through the arterioles so heat is lost by radiation and convection.
2   Arterioles constrict so less heat lost through radiation.
3   Thyroid gland secretes extra thyroxin. Thyroxin increases the metabolic rate which increases heat production.
4   A number of diseases are accompanied by characteristic changes in body temperature; it is a way of monitoring the course of certain diseases and it is a way of a doctor evaluating the efficiency of a treatment.
5   Depends on the stage of hypothermia. Mild: shivering and muddled thinking. Moderate: violent shivering, inability to think, weak pulse, slow and shallow breathing. Severe: shivering stops, difficulty in speaking, little or no breathing. Treatment: call for emergency medical help; monitor breathing – expired air resuscitation if needed, CPR if pulse/heart stops; move person out of the cold or cover head and insulate body from cold ground; replace wet clothing with dry blankets.

6   Hypothermia occurs when the body's core temperature drops below 35 °C. Hyperthermia occurs when core body temperature rises significantly above normal.
7   Mental confusion, headache, muscle cramps and vomiting. May act drunk or combative. Treatment: goal is to lower body temperature quickly – take the person to a cool place, bathe in cool water, cover with cold towels, use a fan to wick heat away from the skin. Patient must also be hydrated – intravenous drip/isotonic drinks. Advanced state – hospitalise.

Diabetes (page 156)

1   Islets of Langerhans
2   Glucagon
3   Insulin
4   α cells stop secreting glucagon; β cells secrete insulin; insulin binds to receptors in liver, fat and muscle cells, so the uptake of glucose by these cells is increased; the use of glucose in respiration is increased and liver cells convert glucose to glycogen which is stored.
5   β cells stop secreting insulin; α cells start to secrete glucagon; less glucose is taken up by the target cells and the rate of use of glucose decreases; gluconeogenesis occurs – amino acids, pyruvate and lactate are converted to glucose for respiration; glycogen is converted to glucose in the liver and released into the blood.
6   In type 1 diabetes the body is unable to produce insulin because the β cells are destroyed by its own immune system, whereas in type 2 the body target cells lose their responsiveness to insulin. Type 1 usually begins in childhood whereas type 2 does not usually start until a person is over the age of 40.
7   Regular check of blood pressure is important because a major cause of death for diabetics is coronary heart disease. Regular examination of the retina is important because in poorly controlled diabetes the tiny blood vessels in the retina are damaged. This may cause blurring and occasionally loss of vision. A kidney function test is important because a diabetic may hold excess sugar in their urine and so be more likely to develop bladder infections such as cystitis. These infections can then spread up to the kidneys. In poorly controlled diabetes, high blood glucose and high blood pressure levels can affect the capillaries in the glomerulus.

Excretion of nitrogenous waste (page 158)

1   To prevent the build up of toxins in the body.
2   Renal artery
3   Cortex, medulla, pelvis. Cortex – outer region; medulla – inner part of the kidney; pelvis – funnel-shaped cavity that receives the urine.
4   Bowman's capsule, proximal convoluted tubule, distal convoluted tubule.
5   Glomerulus
6   U-shaped tube: consists of an ascending limb and a descending limb. It is situated between the proximal and distal convoluted tubules. The loop has a hairpin bend in the renal medulla.

How a kidney works (page 160)

1   To act as a molecular filter – no cells or proteins with a molecular mass greater than 69 000 can get through (they stay in the blood).
2   To increase the surface area for selective reabsorption.
3   The more mitochondria the more energy that can be released in the form of ATP for active uptake.
4   To regulate the chemical composition of the body fluids. The primary function of the kidneys is to maintain a stable internal environment – they do the job of conserving water, salts and electrolytes. All glucose molecules, amino acids and most vitamins are recovered, although the kidneys do not regulate their concentrations. The reabsorption of the ions Na^+, K^+, Cl^-, Ca^{2+} and HCO_3^- occurs at different rates depending on feedback from the body.

5 Active – reabsorption of sodium ions; energy required. Passive – no energy required; reabsorption of water by osmosis.

6 The water potential of our blood is monitored by osmoreceptors (cells) in the hypothalamus of the brain. A fall in water potential of blood flowing through the hypothalamus is detected, resulting in a loss of water by osmosis, so the osmoreceptor cells shrink. This change stimulates neurosensory cells in the hypothalamus to produce and secrete antidiuretic hormone (ADH). ADH passes along the axons of the neurosensory cells which end in the posterior pituitary gland from where ADH is released.

7 The insertion of aquaporins into the collecting duct cell membrane where they open, making the membrane more permeable to water.

More on the kidney (page 162)

1 Dehydration, diabetes insipidus, diabetes mellitus.

2 Diabetes mellitus is the inability to control blood glucose levels, whereas diabetes insipidus is the inability to control the balance of water in the body.

3 High blood pressure; homeostasis.

4 When oxygen levels are low.

5 Acute renal failure can be caused by a serious illness or operation; a sudden loss of large amounts of blood or tissue fluid; a sudden blockage, e.g. a kidney stone; a side effect of some medications or bacterial infection in the kidneys. Chronic renal failure can be caused by a complication of diabetes mellitus; uncontrolled high blood pressure; inflammation affecting the kidney tissue; or be the result of certain inherited diseases, such as polycystic kidney disease. Any two examples for each.

6 By ultrasound or CT scan.

Treatment of kidney disease (page 164)

1 Erythropoietin drugs to curb anaemia; drugs to help control calcium and phosphorus levels; special diet to decrease workload on the kidneys, keep body fluids and minerals in balance, and stop a build-up of toxins; blood pressure control – important in order to prevent or delay the progression of kidney disease; renal dialysis and/or a transplant.

2 To increase the rate of diffusion.

3 To prevent the blood clotting in the dialyser.

4 Haemodialysis – blood is circulated through a machine, containing a membrane, and back into the patient via catheters. Haemodialysis can be carried out in a special unit in a hospital or at home. Peritoneal dialysis – dialysing fluid is circulated into and out of the abdomen using the patient's peritoneal membrane as a filter. Peritoneal dialysis can be done in the patient's home, workplace or almost anywhere.

5 Allows the exchange of substances between the blood and the dialysis fluid. The dialysate contains the correct concentrations of salts, urea and water, so any substances in excess in the blood will diffuse across the membrane into the fluid and vice versa.

6 Blood can be lost as residual blood in the dialyser or blood lines, through frequent blood testing or through a dialyser catheter when blood is discarded routinely at the start of dialysis, so it is important that the patient is given erythropoietin.

7 A better quality of life.

Module 4 – The third age

Effects of ageing on the reproductive system (page 166)

1 In women: chance of pregnancy decreases with age; woman's chance of having a miscarriage increases with age; quality of oocytes decline with age; FSH levels increase with age;

menopause occurs. In men: reduction in the number and motility of sperm produced; reduction in the percentage of normal sperm produced; levels of testosterone reduce; reduction in sex drive; frequency, length and rigidity of erections decline; blood flow in penis is reduced together with the volume of the ejaculate. In both: reduced reproductive capacity. Any six examples.

2 Changes in a woman's physiology are due to a lack of oestrogen and progesterone; levels of cholesterol increase; for the first 5 years after the start of menopause calcium loss from the bones increases resulting in a loss of bone density; the percentage of body fat increases and muscle mass decreases leading to a change in body composition.

3 HRT is prescribed because it increases the declining levels of oestrogen so can relieve symptoms such as hot flushes and sleep disturbances.

4 No need to remember to take a pill every day. Not likely to interfere with other medication.

5 If used for a long time, the risk of breast cancer, endometrial cancer, blood clots and stroke is increased.

Effects of ageing on the nervous system (page 168)

1 A degenerative disease of the nervous system. Possible causes: age – most important known risk factor; family history; severe head or whiplash injuries; smoking, high blood pressure, high cholesterol; a gene which is responsible for production of beta-amyloid protein.

2 Emotional, physical, social and financial burden.

3 Loss of hearing linked to changes in the inner ear. Speech becomes difficult to understand, especially when there is background noise.

4 A cataract is clouding of the normally clear lens. It is treated by surgically removing the cloudy lens and replacing it with an artificial lens.

5 To check for glaucoma.

6 There is a loss of central vision. It affects the ability to read and to see fine detail.

Effects of ageing on skeletal, cardiovascular and respiratory systems (page 170)

1 Bone mass reduces; osteoblast activity declines but osteoclast activity continues at normal level; bones break more easily and do not repair well; vertebrae may collapse; intervertebral discs shrink; incidence of osteoarthritis increases; osteoporosis – bones affected are less dense than normal bone so fractures are more likely. Any four effects.

2 Osteoarthritis – affects the joints due to wearing away of the cartilage, whereas in osteoporosis the bones affected are less dense than normal bone so fractures are more likely.

3 Intervertebral discs shrink – compressed discs and loss of bone mass lead to decrease in body height.

4 This needs to be started in childhood by drinking milk or eating milk-based products so have sufficient calcium and vitamin D; exercise, particularly weight-bearing exercise; not smoking and avoiding excessive intake of alcohol; check home safety to prevent falls; post-menopausal women may need to replace oestrogen by taking HRT.

5 Lung becomes less elastic and therefore less effective. Ribcage does not move as freely.

6 Arteriosclerosis – hardening of the arteries. Atherosclerosis refers to the build-up of fatty deposits called plaques in the walls of the arteries.

7 Coronary arteries narrow restricting blood supply to the cardiac muscles.

Answers to end-of-unit questions

Unit 1 – Molecules, blood and gas exchange

1 (a) Phospholipid bilayer; detail, e.g. heads point out/tails point in; proteins present in it; detail, e.g. description of intrinsic/extrinsic; fluid mosaic; refer to cholesterol; correct reference to glycoproteins/glycolipids; (3)

 (b) **A** osmosis;
 B diffusion;
 C active transport;
 D facilitated diffusion; (4)

 (c) Less ATP/energy available; less enzyme activity, in mitochondria; less K^+ pumped in; K^+ (still) diffuse out; (2)

 (d) (i) HIV/hepatitis, antibodies present; Reject: 'antibodies' alone; complementary/bind, to the antigen; (2)

 (ii) any two from
 If recently given a donation; if received a donation; if anaemic; under age; qualified reference to weight; correct ref. to genetic disorder, e.g. sickle cell/Crohn's; current infection/named infection; recent piercings; recent tattoo; qualified reference to overseas travel; sex with a high-risk group, e.g. drug users; highly promiscuous; pregnant; drug user; certain medications; recent surgery; recent immunisations; AVP; (2)

2 (a) (i) **A** red blood cell/erythrocyte;
 B (squamous/pavement) epithelial; (2)

 (ii) Surfactant reduces cohesive nature of water molecules; reduces surface tension; prevents lining of alveolus sticking together/allows alveoli to expand; Reject: allowing lungs to expand; (2)

 (b) (i) Trace **D** is showing more frequent and deeper breaths; or less frequent and shallower breaths. (1)

 (ii) 26–27; (1)

 (iii) Any two from:
 Use medical grade oxygen; renew hydroxide/soda lime regularly; check for asthma/other respiratory disorder; check for heart conditions; disinfect mouthpiece; Reject these answers: clean; risk assessment; AVP, e.g. check medications, infection, time can only be used if qualified by an explanation such as frequency (2)

 (iv) Oxygen being used (from drum/container); CO_2 being breathed out combines/absorbed (with hydroxide); volume of gas decreases; (2)

3 Method marking points: award a maximum of 5 for method marking points M1–9.

 M1 detail of obtaining blood, e.g. sterilise skin with alcohol;
 M2 dilution of sample;
 M3 method of setting up haemocytometer slide with cover slip;
 M4 description of haemocytometer grid cells A, B and C;
 M5 description of cell count method/number of cells used in count;
 M6 description of treatment for counting overlapping cells/count North/West rule of cell counts for those cells touching or overlapping the grid lines;
 M7 addition of blood sample to slide;
 M8 use of blood pipette/description;
 M9 Dacie's fluid;

Analysing marking points: award a maximum of 4 for marking points A1–5.

 A1 means calculated;
 A2 description of multiplication for $1\,mm^3$ volume;
 A3 reference to dilution to correct blood cell number;
 A4 reference to expected red cell count number;
 A5 results compared; (7)
 Quality and use and organisation of scientific terms: e.g. erythrocytes, haemocytometer, Dacie's fluid, blood pipette, dilution; (1)

4 Polysaccharide; glucose; glycosidic; condensation; water; muscle; energy; respiration; (8)

5 (a) Phagocyte/named phagocyte/macrophage/B cell/T cell/lymphocyte; (1)

 (b) Clots may cause blockages; in coronary arteries; inhibitor changes active site; inhibitor blocks enzyme action; fewer prostaglandins produced; fewer prostaglandins results in reduced clotting; allosteric inhibition; substrate can no longer fit; AVP, e.g. refer to enzyme–substrate complexes, reversible/irreversible, permanent/temporary/competitive inhibitor; AVP; blood flow to heart muscle not restricted/oxygen and nutrients supplied to heart muscle; (5)

 (c) Tissue damage; platelets activated; release of thromboplastin; prothrombin to thrombin; fibrinogen to fibrin; fibrin insoluble; red blood cells caught in mesh/fibres/threads to form a clot; role of calcium ions; AVP, e.g. refer to cascade reaction, clotting factors, fibrin, fibrous protein; Acceptable answer: annotated flow diagram; (4)

 (d) Prevents formation of blood clots in coronary arteries; blood flow maintained, to heart/cardiac, muscle; AVP, e.g. reduces embolisms; reject: thinning of blood (2)

6 (a) Amount of blood pumped from left ventricle; in 1 minute (Acceptable answer: in a given amount of time); CO = HR × SV (cardiac output = heart rate x stroke volume) (2)

 (b) (As exercise intensity increases) blood to muscles increases; (as exercise intensity increases) blood to brain/ skin/decreases; figs to support; more respiration in muscles; more blood with glucose/oxygen; diversion due to vasoconstriction/shunting; AVP; removal of heat/CO_2/lactate, increased ATP production; (3)

 (c) (i) 14.3;
 1 mark for
 12 – 10.5 / 10.5 x 100; (2)

 (ii) Training increases heart muscle size; correct reference to hypertrophy; greater force of contraction of heart; AVP, e.g. Starling's law of the heart; (2)

 (iii) Different morphs/body shapes; different masses; different genders; different genetic factors; different diets; different training regimes; AVP, e.g. smokers/non-smokers, medications, different initial fitness; AVP; Reject: different ages. (2)

Unit 2 – Growth, development and disease

1 **(a)** **(i)** To increase Mary's awareness of the condition; to monitor effect of diet changes; untreated/long-term effects of diabetes can be very serious; such as increased risk of CHD/stroke/blindness/gangrene/neuropathy, shortened lifespan; the results could help Mary control her diabetes with diet and exercise; **(5)**

 (ii) In the morning she will find her fasting blood glucose level; establishes a base line; if this is high there is a real problem; may indicate need for further test, such as blood insulin levels. **(5)**

 (b) Avoid burgers and chips; reduce sugar intake/fizzy drinks; don't eat processed food; eat more fresh fruit and vegetables; eat more fibre/complex polysaccharides; greatly reduce saturated fat intake; use very little salt in diet; **(3)**

2 **(a)** By HIV infection; which attacks B/T cells/macrophages; or by taking immunosuppressant drugs; after a transplant operation; **(2)**

 (b) **(i)** Lymphocytes develop into plasma cells after they have been selected; by being presented with the correct antigen; and selected B cell clones by mitosis; **(2)**

 (ii) Bacterium/virus/protoctist; **(1)**

 (iii) Child receives antibodies from mother; across placenta or in breast milk; this gives passive immunity; to disease mother has been exposed to; these antibodies may last for up to a year while child's immune system is developing; **(3)**

 (c) Quality of life of child; whether to have children whether to tell relatives; refer to gene therapy individuals may not wish to be screened; treatment/cure may not be available; there may be less funding for research into rare genetic diseases; **(4)**

3 **(a)** Each new set of chromosomes moves towards/is pulled; to opposite poles; by spindle threads contracting; **(3)**

 (b) **(i)** the DNA is copied; during interphase; so each chromosome consists of two identical sister chromatids; as these are separated each new nucleus ends up with a full set of chromosomes; that are exactly the same/have the same *alleles*; **(3)**

 (ii) Any two of: growth/repair/asexual reproduction; **(2)**

 (c) **(i)** interphase; **(1)**

 (ii) prophase; **(1)**

 (iii) interphase; **(1)**

 (d) Cell cannot complete the cycle; so does not keep dividing; by mitosis; **(2)**

 (e) It produces haploid gametes; each has half the number of chromosomes/only one set; so when two gametes join at fertilisation; chromosome number restored; diploid zygote; **(3)**

4 **(a)** **( )** ultrasound; (hypodermic) syringe/needle; through abdomen/vagina into womb; sample of chorionic villus tissue/cells; from placenta; **(3)**

 (i) may cause miscarriage (of fetus); may introduce infection; **(1)**

 (iii) Any two of: decision whether to abort fetus or not; decision to have child that they know will have the disease; religious/cultural objection to medical intervention; might result in loss of healthy fetus; is having the test worth the increased risk of miscarriage; **(2)**

 (b) 37/313 × 100; = 11.82%/11.8%/12%;

 (c) **(i)** Turner's syndrome – only 45 chromosomes, 1 X;

 (ii) Klinefelter's syndrome – 47 chromosomes, XXY; **(2)**

 (d) Advantage – amniocentesis has less chance of miscarriage; disadvantage – cannot be done until later in pregnancy so more difficult to opt for termination; **(2)**

5 **(a)** growth rate drops across age range 0–20 years; growth rate drops rapidly initially, up to 3/4 years; growth rate falls slowly within period 3/4 to 11/12 years; growth rate increases/growth spurt from 11/12 years; growth rate drops rapidly between 15 and 18/20 years; growth levels off at 20+; comparative figures to support trends; correct reference to production of growth hormone; growth spurt occurs at puberty; fully developed at 20; **(4)**

 (b) record the height at beginning and end (of the month or year); calculate increase in height; add increases in heights together; divide by the number of children or divide by number of months followed by number of children; **(2)**

 (c) Any two of: carbohydrates/glucose/starch; for energy; lipids; for energy/essential fatty acids/cell membranes/hormones; proteins/amino acids; to make, new protein or type of protein or name of a protein; calcium (ions); to strengthen, bones/teeth; iron; (formation of) red blood cells/haemoglobin; vitamin A; vision; vitamin D; bone strength; phosphorus; to make nucleic acids/DNA/RNA/nucleotides; or ADP/ATP; vitamin C; to make collagen at joints and in blood vessel linings; folic acid; to promote cell division or healthy nervous tissue; iodine; functioning of thyroid/production of thyroxine; **(4)**

6

 (a)

Type of immunity	Gives immediate protection	Gives long-lasting protection
Passive natural	yes	no
Active natural	*no*	*yes*
Passive artificial	*yes*	*no*
Active artificial	*no*	*yes*

 (3)

 (b) **(i)** plasma cell; **(1)**

 (ii) A is variable region; B is constant region; C is disulfide bond; **(3)**

 (iii) The variable region has a specific (3D) shape; that is complementary to the shape of the antigen; this is due to the specific sequence of amino acids/primary structure (of the variable; region); **(2)**

 (c) exposes **antigen** to host cells; refer to **macrophages** and antigen presentation; selection of correct B cell/**clonal selection**; division by **mitosis**;

memory cells and plasma cells; memory cells remain and confer immunity;
example of immunity by this type of vaccination; injection of antibodies; gives immediate protection/immunity; temporary as these antibodies are foreign; example given; (7)
(For clear account including four specialist terms and correct spelling: 1)

7 (a) production of *genetically* identical cells or organisms; (1)
   (b) (i) (stem cells are) unspecialised or undifferentiated; pluripotent/totipotent/omnipotent, or have the potential to, develop/differentiate into many types of cell or tissue; (stem cells) keep dividing; produce large numbers of cells; (2)
       (ii) (culture) *medium*; sterile; (medium must contain) growth factors; (medium must contain) nutrients; suitable temperature/pH; (2)
       (iii) cells becomes adapted, for their function/to carry oxygen; they become differentiated/specialised; genes switch on or off; causing haemoglobin to be made; loss of nucleus; take on, special shape/biconcave disc; (3)
   (c) more tissues/organs available; no need for donors; reduced waiting time; will not cause a strong immune response or be rejected by recipient; recipients won't need to take immunosuppressive drugs for life; less likely to transmit infectious disease; (2)

8 (a) HIV infects cells of the immune system; B/T cells or macrophages; so host is immunocompromised; more susceptible to opportunistic infections; (3)
   (b) Lack of symptoms so difficult to diagnose; no vaccine; virus mutates; changes antigens; no cure; lack of education about cause/methods of spread; lack of money so no free condoms; cultural barriers, reluctance to use condoms; stigma so people may not be tested; spread by unprotected sex; difficult to get people to change behaviour; blood transfusions may be unscreened; may pass from mother to baby; needles for injections may be unsterilised; (7)
        quality of spelling, punctuation and grammar; (1)

9 (a) (i) Disease that spreads across continent/across the world; (1)
       (ii) NO/few individuals in population are immune; no vaccine available; plenty of hosts to infect; need to collect epidemiological data to find out how it spreads/at-risk groups; (2)
   (b) mutation; proto-oncogenes; mitosis; tumour; benign; malignant; (6)

10 (a) Mistletoe cell has a cellulose cell wall; has a chloroplast; has a large permanent vacuole; does not have centrioles; (3)
   (b) 313; 69/100 x 453; (2)
   (c) spindle threads cannot form so no mitosis/DNA polymerase can't be made; (1)
   (d) mutation in genes concerned with cell division; proto-oncogenes to oncogenes; cells fail to undergo apoptosis; keep on dividing by mitosis; may be mutations in tumour suppressor genes; tumour cells may spread; metastasis; malignant;
        T lymphocytes; T cytotoxic/killer T cells; recognise self cell but with changed antigens; insert perforins; and chemicals such as hydrogen peroxide; kill cancerous cell. (7)

Unit 4 – Energy, reproduction and populations

1 (a) Any two of: deoxyribose sugar in DNA, ribose in ATP; one phosphate group in DNA, three in ATP; no pyrophosphate bonds in DNA, two in ATP; AVP, e.g. four different types of nucleotide/named in DNA; (2)
   (b) (i) Denature proteins/enzymes; bonds broken/change in shape of active site; substrate no longer fits; enzymes not able to work at optimum level; AVP, e.g. named bonds; (2)
       (ii) Phosphate group removed from ATP; transferred to/ phosphorylates, other molecule; named example; so energy not lost as heat/energy transferred to other molecule; used as chemical energy; refer to transfer to named action; only a small quantity released as heat/ AW; excess/refer to inefficient coupling; AVP, e.g. detail; (2)
   (c) (i) Chromosomes condense/coil up/AW; homologous chromosomes, pair up/form bivalents; crossing over/ described; chiasmata formation; nuclear membrane begins to break down; spindle begins to form/centrioles move to poles;
           Accept: clearly labelled diagrams (4)
       (ii) Any two of: DNA replication/detail; protein synthesis/ detail; spindle formation; depolymerisation of spindle fibres; spindle contraction; replication of organelles/ named; AVP, e.g. formation of new nuclear membrane; (2)

2 (a) (i) Increase in tolerance to lactate; number of mitochondria; size of mitochondria; enzymes in, Krebs cycle/glycolysis/AW; electron transport molecules; myoglobin; size of muscle fibres; enzymes for fat metabolism; AVP, e.g. number of myofibrils/further detail of points above (4)
       (ii) VO_2 max reached/small amount of aerobic respiration possible/AW; anaerobic respiration; lactate; hydrolysis of ATP; to form ADP + P; ATP resynthesised from CP (creatine phosphate); AVP; (3)
   (b) A myosin;
        B actin;
        C Z line/disc; (3)
   (c) (1) calcium ions, bind to troponin/cause exposure of binding sites on actin; (2) troponin displaces tropomyosin/ troponin–tropomyosin complex, moves/changes shape; (3) myosin heads attach to actin filament; (4) myosin head is an ATPase; (5) myosin head changes position/actin filaments slide past myosin filaments; (6) each movement is 10 nm; (7) myosin detaches from actin filament; (8) ATP causes this release/AW; (9) hydrolysis of ATP/myosin head cocked; (10) ADP + Pi released; (11) myosin head reattaches; (12) process continues provided enough ATP/calcium ions present; (13) AVP, e.g. I band shortens/Z lines closer together/H zone shortens/sarcomere shortens; (7)
        QWC – in the answer, 1,2,3,5,7,9 should appear in this sequence. (1)
   (d) Increased protein (intake); (as source of) specific/essential amino acids; for proteins within muscle fibres; accept named proteins; named mineral qualified; creatine; AVP, e.g. carbohydrate qualified; (3)

3 **(a) (i)** CO_2 dissolves; to form carbonic acid/H_2CO_3 which dissociates; to form H^+; and HCO_3^-; through phospholipid bilayer; diffusion/AW; hydrophilic channels; AVP; (3)

(ii) Decarboxylase;
Found in the link reaction and in Krebs cycle; (1)

(iii) By inhibiting the respiratory centre; stimulated as lungs fill with air so prevents over-stretching/AW; a low expiration to occur/AW; (2)

(b) Hydrogen ions lower the pH of the blood; produced continuously from respiration; enzymes work best at optimum pH; denatured by extremes of pH; (3)

(c) Reduced; energy; gradient; ATPase/ATP synthase; Accept: ATP synthetase (3)

4 **(i)** 21 million; (1)

(ii) 26%;
one mark for
$$\frac{21 \times 10^6}{80 \times 10^6} \times 100$$
1 mark for error carried forward. Allow 1 mark for correct calculation from incorrect number deducted. (2)

(b) For Philippines – or reverse argument: where appropriate for UK
Higher birth rate; less contraception; explained; high infant mortality rate; fewer doctors/nurses/medical personnel; fewer clinics/hospitals/healthcare; rural/isolated communities; civil unrest/war; (internal) migration/displaced persons/refugees/slum conditions; famine/crop failure/poor food supply; malnutrition; poor sanitation/sewage treatment; poor supplies of clean drinking water/treated water; increased risk of infectious disease; named disease (e.g. malaria, TB, HIV, cholera); poor housing/overcrowding; lack of medicines/vaccines/medical technology; AVP; (6)
For linking an explanation to pattern shown on the graph (1)

(c) Extensive: No fertiliser added/manure; recycles nutrients; low yields; no pesticides; little machinery; low density animal husbandry; more sustainable;
Intensive: Fertiliser added; does not recycle nutrients/nutrients lost; high yields; pesticides used; large amounts of machinery; high density animal husbandry; less sustainable; (2)

(d) Grain yield increased; height of stem decreased; biomass increased; use of comparative figures to describe a trend; (2)

5 **(a) (i)** These points may be implied by the way the figures are quoted
Women's fertility 100%/women most fertile at 20/at maximum, from 20 to 30/men reach maximum fertility at 30 years; than female; women infertile/lose fertility at 50/men still 30% of fertility at 80; (3)

(ii) Male matures later; women have all/finite number of oocytes at birth/sperm production continuous in men; oocytes are older/fewer (ovarian) follicles/oocytes; Reject: ova, eggs, gametes oocytes exposed to pollution longer than sperm; AVP, e.g. approaching/AW, the menopause/decrease in oestrogen; male evolved to have time to spread his genes/AW; (3)

(b) (i) As hormones decline/become irregular/lose balance/AW; may release two oocytes/ova per cycle; Reject: eggs refer to age, of gametes; AVP, e.g. more likely to need IVF; hormonal treatment may result in two follicles; Reject: two follicles unqualified (1)

(ii) Identical twins are genetically identical/non-identical twins are not/no more identical than ordinary siblings;

can indicate an environmental effect/AW; other variables are similar/AW, e.g. of variable controlled; shows a genetic effect influencing response/ora; therefore if both identical twins react in the same way/ora; can compare degree of concordance/similarity, between two groups; AVP, e.g. refer to control group (3)

(c) Menstruation ceases; hot flushes; sweats; night sweats; thinning of hair; loss of body hair; development of facial hair; loss of bone density; (3)

6 **(a)** Chlorophyll a; Accept: chlorophyll for one mark as an alternative to chl. a and b
chlorophyll b;
Other possible answers: e.g. xanthophyll, carotene (2)

(b) (i) Thylakoid/lamella/granum; Accept: membranes
Reject: inner membrane (1)

(ii) Must be a comparative statement
Different, reaction centre/form of chlorophyll a/ absorption wavelengths/700 nm (PS1) and 680 nm (PS2)/PS1 mainly on inter-grana lamellae and PS2 mainly on granal lamellae; Reject: different pigments
Accept: cyclic photophosphorylation involves PS1 only;
Accept: PS1 not involved in photolysis/AW; (1)

(c) X and Y = ATP and reduced NADP;
Need both for one mark (1)

(d) **(1)** Occurs in stroma; **(2)** a series of enzyme-controlled reactions; **(3)** carbon dioxide fixed by RuBP; **(4)** carboxylation; **(5)** enzyme is Rubisco; **(6)** (unstable) 6C intermediate; **(7)** forms (two molecules) of GP; **(8)** forms TP; **(9)** using ATP (linked to 8); **(10)** reduction step; **(11)** using reduced NADP; **(12)** refer to either ATP or reduced NADP coming from light-dependent reaction; **(13)** (most of) TP regenerates RuBP; **(14)** rearrangement of carbons to form pentose sugars; **(15)** ATP required for phosphorylation/ ribulose phosphate to RuBP; **(16)** AVP, e.g. TP can be used to form, lipids/amino acids/hexose sugars/suitable named example; (7)
Award if 3, 6, 7, 8 and 13 appear in sequence (1)

Unit 5 – Genetics, control and ageing

1 **(a) (i)** male Nn ; / I^A N I^O n or vice versa
$I^A I^O$;
female nn ; / I^O n I^O n
$I^O I^O$; **A** upper or lower case for ABO superscript (4)

(ii) Autosomal, i.e. not sex-linked; equally distributed between sexes; dominant; found in each generation; locus for nail patella syndrome allele must be on same chromosome as that for blood group A; (3)

(b) Would explain how female 1 inherited the disease, the fact that it is dominant, occurs in every generation, and in both sexes; people with blood group A more likely to have nail patella syndrome as inherited on same chromosome; female 1 and both her parents are blood group A; no cure for nail patella syndrome and it can be a severe handicap; no genetic test but fetus can be screened for blood group A by amniocentesis or chorionic villus sampling; scan could be recommended which would look for other skeletal deformities; as dominant characteristic, the probability of any offspring having nail patella syndrome is 0.5, although as linked to blood group A this probability will be less; go through the options but would not make any decisions for the couple; (8)

Answers to end-of-unit questions

2 (a) Phytohaemagglutinin encourages mitosis to take place; colchicine is added to stop mitosis by preventing the formation of a spindle; dilute salt solution then added so cells will absorb water due to osmosis, causing them to swell and thus separating the chromosomes, which are then stained; the stained chromosomes are photographed, and the individual chromosomes cut out of the photograph and arranged according to size, forming a karyotype; (3)

(b) Chromosomes may be changed in number or shape, and karyotypes are simple to produce; chromosome mutations usually produce physical, i.e. visible, defects in the fetus; gene probes are not available for all mutations; (2)

(c) Any two of: dilemma whether to have tests; people involved may not want to know and should relatives of affected individual be told; knowledge of a positive result may affect their lives; decision about terminating pregnancy, involves dealing with concerns about destroying a life; if the test is carried out there may be implications for an individual's insurance policies and mortgage; fear that a test may give a false positive result or will not show the degree of disability; (4)

3 (a) A mutation is a random or spontaneous change in the genetic material/chromosome/DNA; either an alteration in DNA sequence (gene mutation) or a change in the number or structure of chromosomes (chromosome mutation); (2)

(b) Only affects that individual; a mutation in a germ cell will be passed on and may affect every cell in the offspring; this is because the primary oocyte develops into the gamete; (3)

(c) Addition of a dominant allele will effectively convert the individual into a heterozygote for the gene, producing at least some correct product from the gene; if the disorder resulted from a dominant allele adding another allele would not make any difference; (2)

4 (a) Because of the shortage of organs and the need for a good match, there could be pressure on relatives to donate an organ; pressure on medical services, and could prove to be more expensive than a number of life-saving cheaper operations; belief that it is not right to interfere with nature and religious objections; temptation to risk selling own organs, if short of money; xenotransplants a possibility but could raise more issues about the use of animals; (4)

(b) Antigens on cell surface membrane of all cells, except red blood cells, hence HLA (human leucocyte antigen) system; coded for by six gene loci which are on chromosome 6, the major histocompatibility complex; each of the six loci has many alleles; there must be a close match between donor and recipient. (4)

5 (a) Any two of: allow action potential/impulses to pass to next neurone/muscle; allow transmission in one direction only; allow impulses to travel to/from many neurones, so increasing the range of response; involved in memory/learning; (2)

(b) By diffusion; through channels/calcium gates which open when the impulse arrives; (2)

(c) Pain relief; (1)

(d) As a competitive inhibitor, i.e. *compete* with the substrate for the active site; similar structure to the substrate. Unacceptable answer: inhibitor is the same shape as the substrate; prevents substrate molecules occupying the active site. As a non-competitive inhibitor it may attach to the enzyme at a place other than the active site/allosteric site; this will alter the shape of the enzyme/active site; (3)

6 (a) A = pupil; B = fovea; C = blind spot; D = vitreous humour; (4)

(b) Optic nerve carries action potentials to the brain; retina receives the light stimulus; suspensory ligaments hold lens in place, accommodation; (3)

(c) Pupil response test: in a dark room a light is shone into the eyes (one then the other); there should be a reflex action and both pupils should constrict/react equally; unequal constriction would indicate damage to optic nerve/brain; (3) Blink reflex test: an object is moved towards the eye; should blink/close rapidly; because this is one of last reflexes lost as unconsciousness deepens, if a reflex is absent the patient must be in a coma or suffering from severe brain damage; (3)

(d) Using a Snellen chart (could be described in words or by a diagram); one eye is covered at a time and the chart is viewed at 6 m (20 feet); it is read from top down to the smallest they can see; the smaller the letter that can be read the better the visual acuity; those who cannot read may view pictures; near vision is tested with a reading card – the person is asked to read (blocks of) text of different font/letter, sizes; (4)

(e) A resting potential is maintained (across membrane of rod cell) because there is a negative charge on inside compared to outside; the synaptic bulbs secrete a steady stream of glutamate (a neurotransmitter) which prevents action potentials/depolarisation in bipolar cell; light changes shape of rhodopsin molecule in a rod cell; causes this molecule to straighten – cis-retinal to trans-retinal; causes Na^+ channels to close; the inside of the cell becomes more negative (it is hyperpolarised); the rod cell stops secretion of glutamate, so there is no inhibition of the bipolar cell and it becomes depolarised, creating a generator potential; (6)

7 (a) A = cerebrum; B = medulla oblongata; C = cerebellum; (3)

(b) Traumatic brain injury is the result of a (sudden) blow to the head or if the head suddenly and violently hits an object; causes (physical) damage to the brain; (1)

(c) Patient lies inside a magnet; a computer measures the magnetic fields, in different parts of the brain; haemoglobin contains iron; the effect is smaller when oxyhaemoglobin is present so information about the distribution of oxyhaemoglobin in the brain is obtained; this represents the rate of respiration (damaged areas = less respiration); (4)

(d) Any three of: high blood cholesterol; diet high in saturated fat; smoking; traumatic brain injury/surgery; aneurysm; ageing; genetics, e.g. family history; (3)

(e) A reflex arc is an automatic response and does not involve conscious thought/the brain; pathway involves sensory and motor neurones and the impulse travels via the spinal cord; impulses for pain travelling to the brain have a longer pathway than that to the effector; hence the response is quicker than the feeling of pain; (4)

(f) Insulates the axon, as sodium and potassium ions cannot flow through myelin sheath; nodes of Ranvier are gaps where myelin sheath doesn't cover axon; action potential jumps from one node/gap to the next; salutatory conduction; (5)

(g) Nerve impulse 'short circuits' because there is no (electrical) insulation so ions leak out of the axon; result likely to be paralysis and/or slurred speech, together with poor coordination; (2)

8 (a) No duct and numerous capillaries; hormones secreted directly into blood; (2)

(b) Cause alpha cells to secrete glucagon which converts glycogen into glucose (glycogenolysis); conversion of amino acids/protein/lipid/fat into glucose is also increased (gluconeogenesis); due to negative feedback, insulin production is stopped; (3)

(c) Glucose decreases water potential of blood, so water moves down a water potential gradient from the cells into the blood; this results in large volumes of water/urine being excreted and the cells/body becoming dehydrated causing excessive thirst; enzymes cannot function efficiently; electrolytes are disturbed which is particularly damaging in the brain, resulting in a coma (becoming unconscious) excess glucose is converted to fat, so breath smells of pear drops; and there is a danger of atheroma forming in coronary arteries (CHD); risk of gangrene in extremities; excess glucose crystallises in the lens which can denature the lens; (6)

9 (a) High pressure is needed in the glomerulus for ultrafiltration; (2)

(b) Glomerular filtrate contains all materials in blood with molecular mass less than 65 000–69 000; in the proximal convoluted tubule (PCT), glucose, amino acids, vitamins and some ions are reabsorbed; sodium ions are actively transported out of the PCT; diffuse into the cells from the filtrate via cotransporter proteins – cotransport/facilitated diffusion occurs as glucose moves into cells with the sodium ions; cells of the tubule have microvilli, which give a large surface area of contact with the filtrate, and many mitochondria; these provide ATP energy for active transport; movement of ions/glucose/solutes sets up a water potential gradient so water moves out by osmosis, making filtrate more concentrated (65% of the water is reabsorbed here); some urea also diffuses out of the filtrate in the PCT; (8)

(c) If blood volume increases there are fewer red blood cells per unit volume; therefore less oxygen carried to kidney cells; (1)

(d) Injecting RhEPO: more red blood cells to carry oxygen for aerobic respiration so more ATP for muscle contraction; (1)

(e) RhEPO is genetically engineered. The gene for RhEPO is inserted into a plasmid and the bacterium then produces RhEPO; (2)

10 (a) Core temperature falls below 35 °C; sleepy/very relaxed/ disorientated because drop in temperature reduces kinetic energy of reactants/enzymes causing enzyme-controlled reactions to slow; metabolism therefore slows; (3)

(b) Wrapping patient in blanket/survival sack; supplying hot water bottle; providing body heat; giving warm IV (intravenous) fluid/warm drink; no hot drinks should be given until patient is warmer; needs to warm up slowly to avoid risk of heart attack; extremities should not be rubbed; patient should not be given alcohol or glucose; should be kept mentally active by talking to them; (3)

(c) Peripheral temperature is affected by the environmental temperature; (1)

(d) If the body is too cold: (smooth) muscle in walls of arterioles (to skin) will contract (vasoconstriction); diverts blood from the skin surface so less heat is lost due to radiation; erector muscles contract which causes hairs on the skin to rise, trapping an insulating layer of (warm) air; we shiver due to contraction of muscles and this produces heat (energy) from the increased respiration of the muscles; (3)
If the body is too hot: the sweat glands secrete sweat; as evaporates uses heat in the skin to provide the energy (water has a high latent heat of evaporation), and so removes heat (from the blood); arterioles supplying blood to the skin dilate allowing more blood to flow close to the surface of the skin (vasodilation); so heat is lost from the blood by conduction/convection/radiation; (3)

11 (a) As hormones decline/become irregular; may release two oocytes/ova per cycle. Unacceptable answer: eggs refer to age of gametes; zygote/blastula, may divide into two; AVP (arginine vasopressin); e.g. more likely to need IVF (in vitro fertilisation), hormonal treatment may result in two follicles; (1)

(b) Identical twins are genetically identical, non-identical twins are no more identical than ordinary siblings; can indicate an environmental effect; other variables are similar; therefore if both identical twins react in the same way; shows a genetic effect influencing response; can compare the similarity, between two groups; (3)

(c) Menstruation becomes irregular/stops. Unacceptable answer: bleeding ovulation becomes irregular/stops; FSH increases/ oestrogen decreases; dry skin/vagina; loss of bone calcium/ osteoporosis; mood swings; night sweats/hot flushes; (3)

(d) Contains oestrogen/oestrogen and progesterone; stops mobilisation of bone calcium/decrease in bone density; (2)

(e) Steroids/fat soluble; in lipoprotein/phospholipid; will diffuse concentration gradient; through differentially permeable membrane; (2)

(f) Cyclical: pills taken in varying combinations; for up to 25 days/3 weeks; may be oestrogen continuous, progestin 10–14 days; then get withdrawal bleeding/'period'; (2)
Continuous: fewer side effects/lower dose; attached to skin/ skin patches/implant; replaced twice a week; if under skin need replacing after 3 months; oestrogen and progesterone tablets taken continuously; avoids hormone surges; (2)

12 (a) Any two of: genetic (potential)/family history; head injury (associated with unconsciousness/(severe) blow; alcohol abuse; aluminium; smoking; HRT; high cholesterol; (2)

(b) Decrease in sensory perception/reduction of named sense: increased pressure in eye/glaucoma; macular degeneration/ described;
lens hardens/less elastic/goes cloudy; cataract;
hair cells in cochlea degenerate; eardrum loses elasticity;
slower nerve conduction; loss of sensation in finger/toe tips; (4)

(c) Burden on carer; reduction of economic input from carer; increased pressure on welfare/NHS services; increase in costs to NHS/described; increase in need for pensions; balance between those contributing to state funds and those using them shifts; increase in need for residential/respite care/homes/carers; increased incentive to find cures for degenerative diseases/named; increased accessibility needed (to shops/other named area); refer to ageism; experience of older people is valuable; pressure or environment (use of resources such as oil, gas; carbon footprint, etc); (6)

(d) Poor nutrition; protein is needed to make antibodies; poor gas exchange; hypothermia – poor temperature control; (2)

(e) Any one of: avoid contact with infected people; avoid crowded places; be vaccinated against flu and pneumonia; keep warm; (1)

13 (a) Reduction in acetylcholine secretion; due to deficiency in choline acetyltransferase; decrease in cytochrome oxidase; neurofibrillar tangles; microtubules/tau (protein); in nerve cell bodies; beta-amyloid protein; forms plaques; between nerve cells; increased level of beta-amyloid 42; plaques in blood vessels/meninges; decrease in brain mass; neurones die/lost; dendrites atrophy/shorter/fewer branches; fewer synapses; evidence of head injury, deficiency of folate; (7)

(b) Does not show on (CAT/MRI) scans; can only see brain shrinkage/increase in size of ventricle; could be any type of brain damage/dementia; observe behaviour/description of behaviour; examination of brain tissue only after death; (2)

199

Index

Index